Eighth Symposium on Large TPCs for Low Energy Rare Event Detection 2016

Journal of Physics: Conference Series Volume 1029

Paris, France
1 – 2 December 2016

ISBN: 978-1-5108-6489-4
ISSN: 1742-6588

Printed from e-media with permission by:

Curran Associates, Inc.
57 Morehouse Lane
Red Hook, NY 12571

Some format issues inherent in the e-media version may also appear in this print version.

Printed by Curran Associates, Inc. (2018)

Institute of Physics
Dirac House, Temple Back
Bristol BS1 6BE UK

Phone: 44 1 17 929 7481
Fax: 44 1 17 920 0979

techtracking@iop.org

Additional copies of this publication are available from:

Curran Associates, Inc.
57 Morehouse Lane
Red Hook, NY 12571 USA
Phone: 845-758-0400
Fax: 845-758-2633
Email: curran@proceedings.com
Web: www.proceedings.com

Table of contents

Volume 1029

Eighth Symposium on Large TPCs for Low Energy Rare Event Detection

1–2 December 2016, Paris, France

Accepted papers received: 9 May 2018
Published online: 11 June 2018

Preface

Papers

IOP Conf. Series: Journal of Physics: Conf. Series **1029** (2018) 011001 doi:10.1088/1742-6596/1029/1/011001

The eight "International Symposium on Large TPCs for Low-Energy Rare Event Detection", took place in Paris between the 5th and 7th of December 2016 at the Institute of Astroparticle Physics (APC) campus - Paris Diderot University.

As usual the conference was organized just before Christmas, which seems to be convenient for most of the people and occurs every two years with almost 120 participants attending.

The scientific program consisted of plenary sessions including the following topics with theoretical and experimental lectures:

- Low energy neutrino physics
- Neutrinoless double beta decay process
- Dark matter searches
- Axion and especially solar axion searches
- Space experiments and gamma-ray polarimetry
- New detector R&D and future experiments

Many people contributed to the success of the conference, but we would particularly like to thank the University of Paris Diderot for providing the nice Buffon auditorium and infrastructure. We also acknowledge the valuable support of CNRS/IN2P3, APC, CEA/Irfu and the University of Zaragoza. We want to thank also the conveners of the sessions, who allowed for a smooth running of the meeting.

The organizers

IOP Conf. Series: Journal of Physics: Conf. Series **1029** (2018) 011002 doi:10.1088/1742-6596/1029/1/011002

Peer review statement

All papers published in this volume of *Journal of Physics: Conference Series* have been peer reviewed through processes administered by the proceedings Editors. Reviews were conducted by expert referees to the professional and scientific standards expected of a proceedings journal published by IOP Publishing.

IOP Conf. Series: Journal of Physics: Conf. Series **1029** (2018) 012001 doi:10.1088/1742-6596/1029/1/012001

Improved Detection of Supernovae with the IceCube Observatory

Lutz Köpke for the IceCube Collaboration [1]

Abstract. The IceCube neutrino telescope monitors one cubic kilometer of deep Antarctic ice by detecting Cherenkov photons emitted from charged secondaries produced when neutrinos interact in the ice. The geometry of the detector, which comprises a lattice of 5160 photomultipliers, is optimized for the detection of neutrinos above 100 GeV. However, at subfreezing ice temperatures, dark noise rates are low enough that a high flux of MeV neutrinos streaming through the detector may be recognized by a collective rate enhancement in all photomultipliers. This method can be used to search for the signal of core collapse supernovae, providing sensitivity competitive to Mton neutrino detectors to a supernova in our Galaxy. An online data acquisition system dedicated to supernova detection has been running for several years, but its shortcomings include limited sampling frequency and the fact that the burst energy and direction cannot be reconstructed. A recently developed offline data acquisition system allows IceCube to buffer *all* registered photons in the detector in case of an alert with low probability to be erroneous. By analyzing such data offline, a precision determination of the burst onset time and the characteristics of rapidly varying fluxes, as well as estimates of the average neutrino energies may be obtained. For supernovae ending in a black hole, the IceCube data can also be used to determine the direction of the burst.

1. Introduction

The IceCube neutrino telescope [1–4], embedded in the clear east antarctic ice sheet, is uniquely suited to monitor our Galaxy for supernovae due to its 1 km^3 size. Deployed at depths between 1450 – 2450 m, the detector is partly shielded from cosmic ray muons; in the inert and $-43\,°C$ to $-20\,°C$ cold ice, the dark rates of IceCube's 10-inch diameter photomultipliers average around 540 Hz. While designed to reconstruct interactions of neutrinos with hundreds of GeV energy, such a low dark rate in conjunction with a vast amount of interactions in the instrumented volume allows for the detection of $\mathcal{O}(10\,\mathrm{MeV})$ neutrinos from galactic core collapse supernovae. The inverse beta process $\bar{\nu}_e + p \rightarrow e^+ + n$ dominates supernova neutrino interactions in the ice, leading to positron tracks of about $0.6\,\mathrm{cm} \cdot E_\nu/\,\mathrm{MeV}$ length which radiate $\approx 180 \cdot E_{e^+}/\mathrm{MeV}$ Cherenkov photons in the 300 – 600 nm wavelength range. Due to the approximate E_ν^2 dependence of the cross section at supernova neutrino energies and the linear energy dependence of the positron track length, the light yield per neutrino interaction scales with E_ν^3. Photons travel long distances in the clear ice such that each light sensor effectively monitors several hundred cubic meters. In more than 98% of the cases, only a single photon from each interaction reaches one of the light sensors because they are vertically (horizontally) separated by roughly 17 m (125 m).

[1] `http://icecube.wisc.edu`

The DeepCore subdetector [5], equipped with a denser array of high efficiency photomultipliers, provides higher detection and coincidence probabilities.

The effect of a high neutrino flux brightening the detector will be clearly seen in the collective rate increase averaged over many sensors. IceCube is sensitive around the clock with 99.7% uptime and is the most precise detector for analyzing the neutrino light curve of close supernovae [6]. Indeed, for a supernova at the center of our galaxy, IceCube will remain competitive even in case the proposed Hyper-Kamiokande [7, 8] detector is realized.

However, the statistical interpretation of the IceCube data does not allow us to reconstruct individual neutrino interactions and estimate their energy, origin and type. Furthermore, the standard supernova data acquisition is based on count rates of individual optical modules stored in 1.6384 ms time bins, which limit the timing accuracy. Additionally, an artificial deadtime of 250 μs, incorporated in the data acquisition system to cut the dark rate in half, has a detrimental effect in the detection capabilities of very close supernovae.

Therefore an improved readout system has been developed to additionally buffer and extract the full photomultiplier raw data stream around supernova candidate triggers (HitSpooling) [9, 10]. In this paper, we will introduce the HitSpool system and discuss the following examples that build on this improvement:

- the application of sophisticated techniques to identify noise hits or hits associated with cosmic ray muons that do not trigger the IceCube simple majority trigger [10, 11],
- the estimation of the average neutrino energy by the use of coincidence probabilities [12, 13],
- a directional measurement by internal triangulation in the case a signal changes abruptly and the possibility to measure an absolute neutrino mass measurement in the case of core collapse supernovae ending up as a black hole [11], and
- the study of standing wave shock signatures in IceCube and gravitational wave detectors.

Since 2009, IceCube has been sending real-time datagrams to the Supernova Early Warning System (SNEWS) [16] when detecting supernova candidate events. A similar alert system, with substantially lower thresholds to trigger gravitational wave observations, is under preparation [17–19].

2. Technical implementation of the HitSpooling system

Fig. 1 contains a logical diagram of the path of data in IceCube and its connection to the supernova data acquisition system. The 5160 digital optical modules (DOMs) in IceCube, containing photomultipliers and readout electronics, are deployed on 86 so-called strings with 60 DOMs per string. The hit stream from each DOM is buffered in a dedicated industrial computer (DOMHub), where the hits are chronologically ordered and packaged by the StringHub component into a form suitable for the application of software-based triggers and event builders. A copy of the raw data is also written to a circular buffer on disk (HitSpool). Old data are overwritten after a spooling cycle of about 72 hours or longer, depending on the storage capacity of the DOMHub.

In the event of a significant transient, a request is sent to the HitSpool system to store hits within a configurable time interval. For supernova candidates, a period of 30 s before

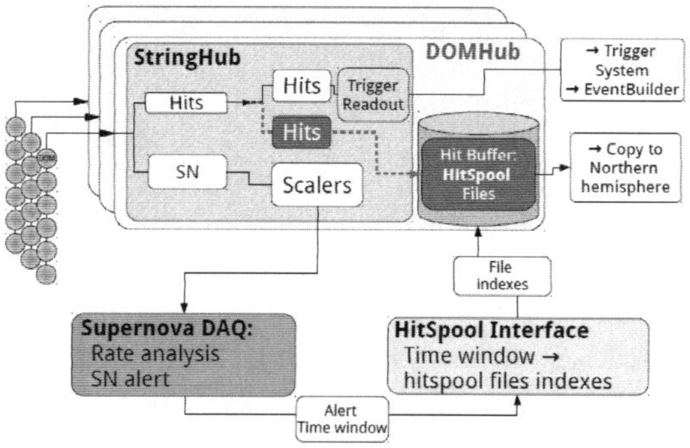

Figure 1: Block diagram of surface data acquisition components involved in the HitSpool data stream (blue dashed line). Hits are copied to the DOMHubs hard disk from where files may be selected and transmitted to the Northern hemisphere via satellite.

to 60 s after a SNEWS trigger is stored. The supernova system performs HitSpool requests on average twice per month. Such data – buffered at an early stage of the data acquisition – will be available in the unlikely case that the data acquisition may fail (i.e. in the case of an extremely close supernova which could exhaust the system). The fully automatic hitspooling system (see data flow scheme in Fig. 1) has been working reliably for several years. HitSpool processing includes the automatic transfer of data to the North and the subsequent monitoring and pre-processing.

3. Examples for recent improvements in IceCube's supernova detection

In this section, we provide four examples that demonstrate how the HitSpooling capability extends IceCube's physics capabilities either by allowing for the exploitation of coincidences between optical modules or by a much improved temporal precision.

3.1. Cosmic ray muon background

Atmospheric muons require an energy of ≈ 400 GeV to reach the detector and ≈ 550 GeV to trigger the IceCube 8-fold majority trigger. Hits from muon tracks that fail this trigger requirement or that clip the corners of the detector are mostly found in the upper or outer layers of the detector (see Fig. 2).

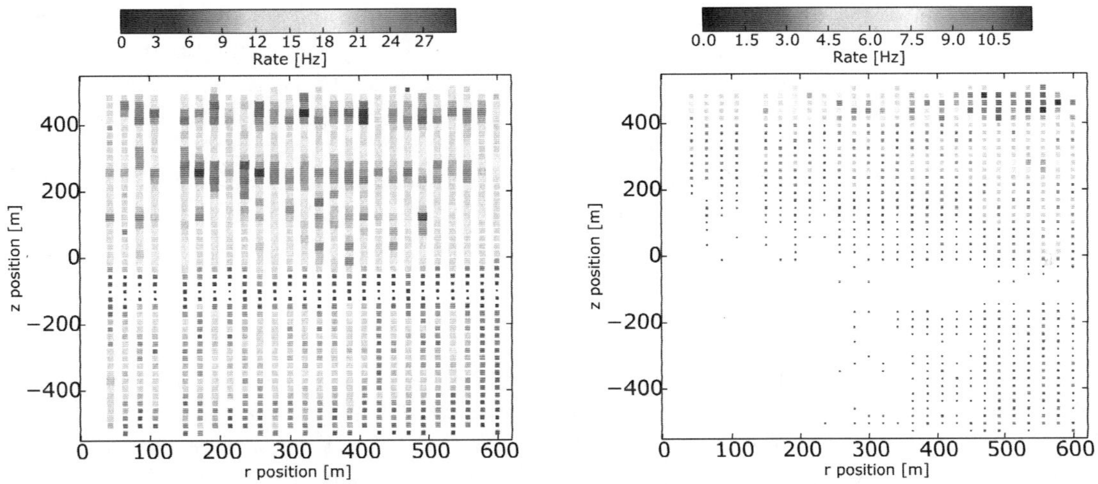

Figure 2: Simulated muon induced hit rates as function of horizontal position and radius w.r.t. to the detector center at 1950 m depth. *Left:* all hits associated with muons triggering the detector. *Right:* hits associated with muons not meeting any trigger condition.

Dust layers in the ice, particularly the one ≈ 100 m below the detector center, absorb light. This effect and the limited muon range lead to a depth-dependent cosmic muon induced rate between ≈ 3 Hz and ≈ 30 Hz. The fraction of such hits in each DOM is small compared to the dark rate. However, space and time correlated hits from muons lead to non-Gaussian tails that broaden substantially the significance distribution of the excess count per sensor. In fact, 99.9% of false positive triggers are due to statistical agglomerations of cosmic ray muon induced hits (see Fig. 3).

To correct for the effect of muons, the number of locally coincident particle induced hits counted by IceCube's majority trigger is transmitted to the supernova data acquisition system and is used to statistically subtract hits associated to cosmic muon induced tracks [11]. The

IOP Conf. Series: Journal of Physics: Conf. Series **1029** (2018) 012001 doi:10.1088/1742-6596/1029/1/012001

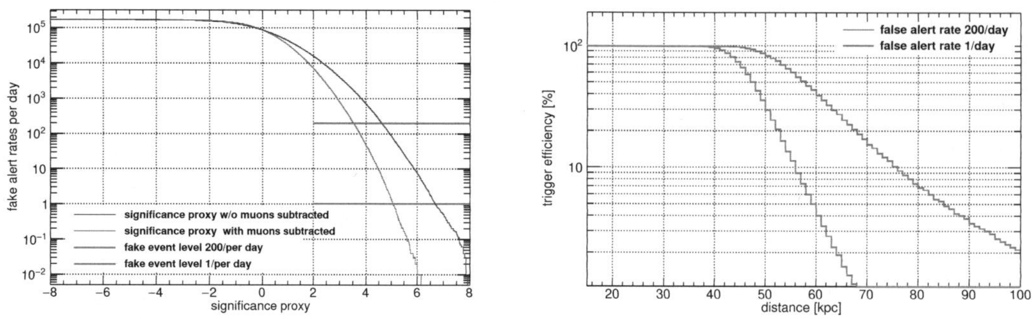

Figure 3: Left: Fake alert rate vs. significance proxy; cosmic ray muon subtraction reduces the rate significantly at large significances. Right: trigger efficiency as function of distance for a high alert rate (200 per day in purple) and one alert per day (green).

method has allowed us to lower the alarm thresholds and reduce false-positive triggers. The muon rate correction has enabled us to improve the SNEWS alarm efficiency for potential supernovae in the Large Magellanic Cloud from merely 12% to 82%, while meeting the SNEWS requirement that alarms are sent with a frequency of less than one alarm per 14 days.

Simulations show that $\approx$50% of all atmospheric muons crossing the detector produce a simple majority trigger. The remaining sub-threshold muons can be identified using algorithms developed to identify hits clustering in time and space combined with an investigation of the angular multiplicity of tracks formed by each hit pair in the cluster. Thereby it is possible to recover up to 45 % of all sub-threshold muon hits, limited only by the requirement of at least 4 hits to define a cluster in the algorithms. Applying these techniques to experimental data results in a $\sim$ 3 % decrease of the total hit rate and reduces the fraction of false positive SNEWS alarms by an additional 45% [10, 11].

3.2. Determination of the average neutrino energy

It has been shown [12] that the average neutrino energies can be reconstructed from a comparison of single rates with coincident rates in the detector. The probability of photons from a single interaction to reach different optical modules increases with the energy of the neutrino. For each optical module added in coincidence, another factor proportional to the neutrino energy is added. This sensitivity of coincident hits to the neutrino energy has been recognized in [13] and was elaborated with a full Monte-Carlo simulation based on Geant4 with custom photon tracking. The uncorrelated background is greatly reduced by the use of a small coincidence gate of 150 ns for IceCube and 100 ns for DeepCore. The correlated background from atmospheric muons has been reduced by a multiplicity cut with further improvements possible (see section 3.1).

The relative precision of the energy estimate is depicted in Fig. 4 as function of distance and average energy. To be most conservative, flux, energy and spectral shape were taken from the collapse of an O-Ne-Mg 8.8 $M_\odot$ progenitor star [14], the lowest mass progenitor known to undergo a core collapse. Systematic uncertainties have not yet been assessed; however, they are reduced as the ratio of the coincidence rate to the singles rate is used in the method.

Multiple coincident hit conditions show distinct differences in the neutrino energy dependence. These can be used to characterize the spectral shape of the neutrino emission, in addition to the average energy. Assuming a neutrino spectral shape parametrized with three parameters, the luminosity L_ν, the average neutrino energy $\langle E_{\bar{\nu}_e} \rangle$ and a shape parameter α [15] and using single, double and triple coincident hit modes in a χ^2 fit, $\langle E_{\bar{\nu}_e} \rangle$ and α can be extracted simultaneously.

A scan of the parameter space for a simulated supernova at 10 kpc distance is shown in Fig. 4

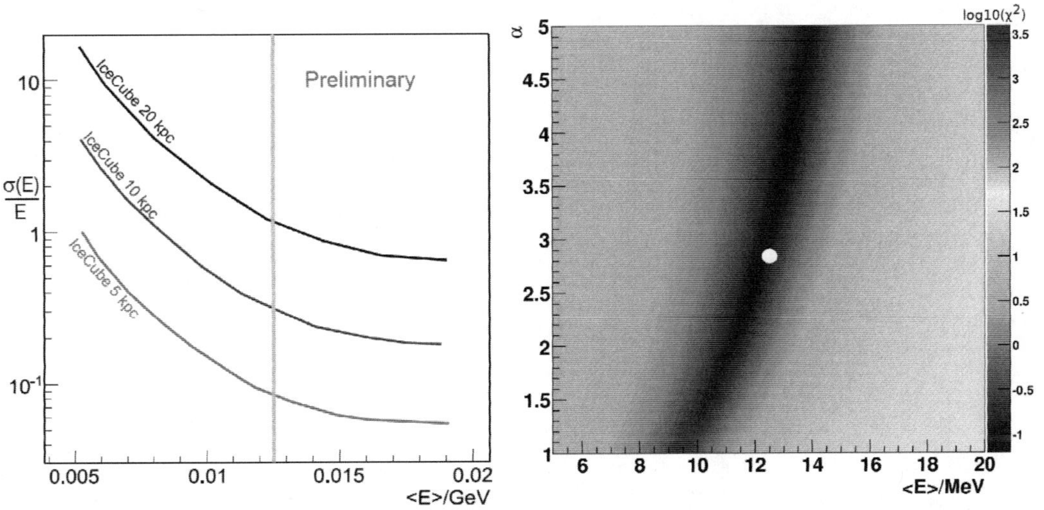

Figure 4: Left: Comparison of the precision on the determination of the average neutrino energy of a supernova at 20 (black), 10 (blue) and 5 (red) kpc distance. Right: Combined determination of the average neutrino energy and the spectral shape parameter α for IceCube using a χ^2 method. The input value ($\langle E_{\bar{\nu}_e} \rangle = 12.6$ MeV, $\alpha = 2.84$) is denoted by a white circle.

as function of average energy and α for IceCube. The color scale indicates the $\log_{10}(\chi^2)$ value for the parameter values tested. Note that α and $\langle E_{\bar{\nu}_e} \rangle$ are almost degenerate.

3.3. Triangulation of the supernova direction and absolute neutrino mass

For neutrino energies between 10-20 MeV, the direction of the emerging positron is essentially uncorrelated with the incoming neutrino direction. Therefore it is a challenge to measure the supernova direction with IceCube. A triangulation between distant detectors has been proposed [20] as one possibility. Another option is to determine the direction from the temporal hit pattern seen in the cubic-kilometer detector when a neutrino wavefront changes its intensity abruptly. The detector crossing lasts only several microseconds, which sets the time-scale for abrupt flux changes to be suitable.

One example is the formation of a black hole following a core collapse of a supermassive star. In case the protoneutron star forms a black hole, the neutrino flux should cease almost immediately, once the Schwarzschild condition is met. However, since the neutrino energies follow a spectrum, non-vanishing neutrino masses lead to a smearing of the time arrival distribution (see Fig. 5) that can be used to determine the absolute neutrino mass [21]. Note that – in the case of rapidly rotating progenitors – the neutrino fluxes may also gradually decrease as more and more matter in the star approaches the event horizon and the gravitational redshift becomes extremely strong [22] (see Fig. 5). We first adopt the optimistic scenario that the smearing due to the black hole dynamics is negligible.

With an unbinned likelihood analysis, simulations show that a reasonable directional resolution can be achieved for nearby supernovae and low neutrino masses. Fig. 5 shows the resolution as function of distance assuming rates and energies according to [23] and an abrupt decrease of the neutrino luminosity at the black hole surface. When using the assumption of a gradual decrease of neutrino luminosity [22] instead, one finds that directions can only be determined for very close supernovae. As the time delay of a neutrino with mass m_i and energy E_ν at distance D, $\Delta t = 0.515 \cdot \frac{{m_i}^2}{\mathrm{eV}^2} \frac{\mathrm{MeV}^2}{E_\nu^2} \cdot \frac{D}{10\mathrm{kpc}}$, is proportional to the squared neutrino mass, the method has only little sensitivity for current cosmological neutrino mass limits. However,

IOP Conf. Series: Journal of Physics: Conf. Series **1029** (2018) 012001 doi:10.1088/1742-6596/1029/1/012001

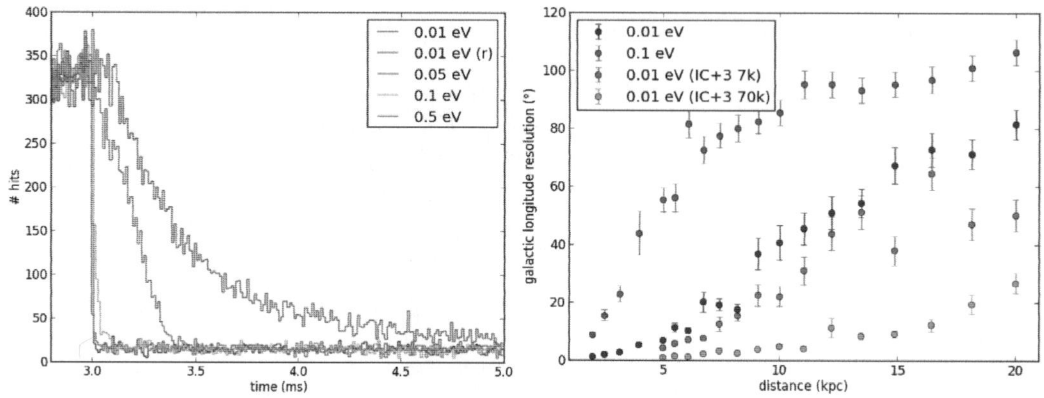

Figure 5: Left: Effect of neutrino mass on the observed neutrino flux for abrupt black hole creation. The green line shows the effect of a gradual decrease of the neutrino luminosity due to rapidly rotating progenitor ($m_\nu = 0.01$ eV). Right: Resolution in galactic longitude achievable by triangulation within IceCube's geometry for a supernova with abrupt black hole formation for two assumptions on the mass of the lightest neutrino (0.01 eV and 0.1 eV, respectively) as well as for a hypothetical detector extension with 3 additional IceCube strings deployed symmetrically around IceCube's center at 7 and 70 km distance, respectively.

sterile neutrinos may make an imprint through their mixing with ordinary neutrinos. Neglecting standard neutrino masses and assuming only one sterile neutrino with mass m_4, the $\bar{\nu}_e$ flux $F_{\bar{\nu}_e}(E_\nu, t)$ at time t after creation is given by

$$F_{\bar{\nu}_e}(E_\nu, t) \approx |U_{e1}|^2 F^0_{\bar{\nu}_e}(E_\nu, t) + |U_{ei}|^2 F^0_{\bar{\nu}_x}(E_\nu, t) + |\widetilde{U_{e4}}|^2 F^0_{\bar{\nu}_x}\left(E_\nu, t - \frac{D}{2c}(\frac{m_4{}^2}{E_\nu})\right) \quad ,$$

with i = 2, 3 for inverted and normal hierarchy, respectively [24]. U_{ei} denote the entries in the upper row of the PNMS matrix. Note that $|\widetilde{U_{e4}}|^2$ should be regarded of as a free parameter since matter and neutrino coherence effects in supernovae, which are hard to predict, may alter the coupling [25]. First simulations with an unbinned likelihood fit show that a sterile neutrino mass $m_4 = 2$eV can be well reproduced for $|\widetilde{U_{e4}}|^2$ larger than 0.5, for an instantaneous black hole forming core collapse at 10 kpc distance. For this analysis, $\mathcal{O}(10\ \mu s)$ timing resolution is essential.

3.4. Supernovae and gravitational waves

While the current sensitivity of Advanced LIGO is limited to supernovae within a few kpc (see e.g. [26]), technological improvements will enlarge the supernova detection range of gravitational wave detectors to that of current neutrino detectors. Lacking observations, there is a strong variation in modeling the core collapse and the occurrence of asymmetries that lead to gravitational wave emission. In case the progenitor star is rotating [27–29], the bounce of the iron core leads to a pronounced spike at bounce time and a subsequent damped ring down with kHz frequencies. Generally, non-axisymmetric rotational instabilities such as standing accretion shock instabilities (SASI), will also result in the emergence of a quadrupole mass-moment during the collapse and thereby leave their imprints on the gravitational wave signal.

The possibility to search for time correlations between neutrino and gravitational wave data was investigated using recent simulations from [26, 30]. Fig. 5 (left) shows the result of a Monte Carlo simulation, both for the neutrino signal in IceCube (normal hierarchy assumed, distance 3 kpc) and a gravitational wave signal at the source (no noise and detector simulation added). The neutrino signal shows a clear oscillatory pattern 150-250 ms after the burst onset. At first sight, no obvious correlations of this rapidly varying signal component are visible in the gravitational wave time series.

It turns out, however, that the effect of standing wave accretion shocks is present in both signals. The Fourier spectra in Fig. 6 (middle) show that the gravitational wave signal, as expected, ocurs with twice the frequency as the neutrino signal. It is also apparent that lower and higher frequencies with larger amplitude contribute to the gravitational wave signal as well. When performing a sliding window cross correlation analysis between the second harmonic in the neutrino frequency spectrum with the gravitational wave frequency signal in 50 ms time intervals, a clear enhancement emerges around 80 Hz (neutrinos) and 160 Hz (gravitational waves), respectively (Fig. 6 (right), with advanced LIGO design detector noise included). This correlation becomes much more apparent at closer distances or when assuming a future gravitational wave detector with lower noise. The higher frequencies seen in gravitational waves do not seem to have obvious counterparts in the simulated neutrino data.

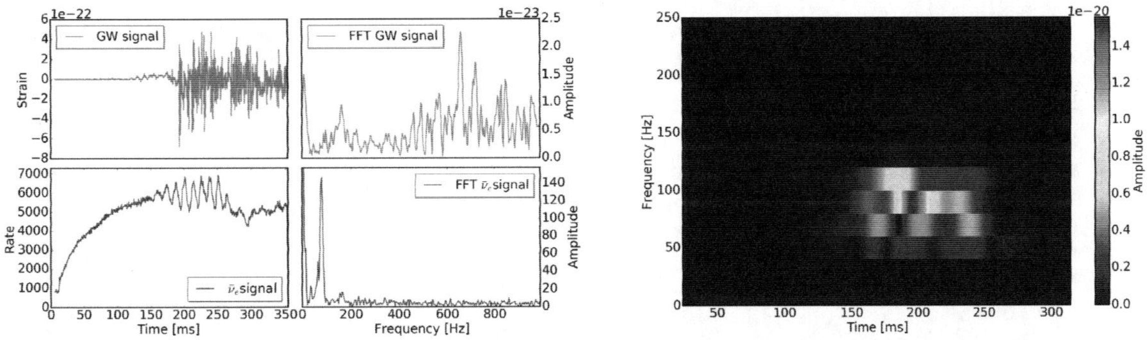

Figure 6: Left bottom: Expected neutrino signal at 3 kpc distance in IceCube [30] (IceCube Monte Carlo), left top: strain amplitude $A+$ of the gravitational wave signal at the source [26], warping space time along the vertical and horizontal directions; middle top and bottom: corresponding Fourier spectra. Right: a spectrogram of the cross correlation between neutrino signal frequency and the second harmonic of the gravitational wave signal with aLIGO design detector noise included.

The possibility to detect oscillations in the 800 Hz range during the first 30 ms for various rotational speeds for rotating progenitors with 12 $M_\odot$ and 40 $M_\odot$ [29] was also investigated. The small signal will only be analyzable for supernovae that are as close as 1 kpc or less [31].

4. Conclusions and Outlook

The IceCube observatory provides the world's best statistical accuracy for the neutrino flux of supernovae in our galaxy with a round-the-clock uptime of currently 99.7%. However, energies and directions of individual neutrinos can not be determined due to the optical sensor dark rates and – to a lesser extent – due to cosmic muons passing the detector. Therefore it is very important to understand the dark rate and cosmic muon characteristics. Average neutrino energies can be determined statistically when analyzing the timing information of all hits in the detector. In the future, the sensitivity to the absolute neutrino mass will be assessed, the energy determination will be improved, and the idea of analyzing $\mathcal{O}(10)$ s long bursts will be extended to shorter potential signals. To improve the sensitivity of the trigger to a wider range of models as well as unusual hadronic physics or physics beyond the Standard Model, we are implementing a time-domain search using the Bayesian Blocks algorithm [32]. This technique allows the data themselves to determine the natural timescale of excess counts above background.

Acknowledgements
We thank Irene Tamborra, Haakon Andresen and Alexander Summa for their helpfulness and providing access to their simulation data and appreciate suggestions by Sarah Gossan on the implementation of gravitational wave detector noise.

References
[1] R. Abbasi et al. (IceCube Collaboration), *Nucl. Instrum. Meth.* **A601** (2009) 294.
[2] R. Abbasi et al. (IceCube Collaboration), *Nucl. Instrum. Meth.* **A618** (2010) 139.
[3] M.G. Aartsen et al. (IceCube Collaboration), *JINST* **12** (2017) P03012.
[4] M.G. Aartsen et al. (IceCube Collaboration), `arXiv:astro-ph/1612.06028`.
[5] R. Abbasi et al. (IceCube Collaboration), *Astropart. Phys. 35* **10** (2012) 615.
[6] R. Abbasi et al. (IceCube Collaboration), *A&A* **535** (2011) A109.
[7] K. Abe et al., Letter of Intent: The Hyper-Kamiokande Experiment, `arXiv:hep-ex/1109.3262v1`.
[8] K. Abe et al., Hyper-Kamiokande Design Report, `https://lib-extopc.kek.jp/preprints/PDF/2016/1627/1627021.pdf`.
[9] D. Heereman, V. Baum, G. Kroll, and B. Riedel (IceCube Collaboration), contribution to the 33rd International Cosmic Ray Conference `arXiv:astro-ph/1309.7008` and paper 444, *Braz. J. Phys.* **44** (2014) no.5, 415.
[10] D. Heereman von Zuydtwyck, PhD thesis, Université libre de Bruxelles, 2015.`http://hdl.handle.net/2013/ULB-DIPOT:oai:dipot.ulb.ac.be:2013/209179`.
[11] V.Baum et al., *PoS (ICRC2015)* 1096.
[12] R. Bruijn, *Nucl. Phys. B Proc. Supp.* **237-238** (2013) 94.
[13] M. Salathe, M. Ribordy and L. Demirors, *Astropart. Phys.* **35** (2012) 485.
[14] L. Hüdepohl, B. Müller, H.-T. Janka, A. Marek, and G. G. Raffelt, *Phys. Rev. Lett.* **104** (2010) 251101.
[15] M. T. Keil, G. G. Raffelt, and H.-T. Janka, *Astrophys. J.* **590** (2003) 971.
[16] P. Antonioli et al., *New J. Phys.* **6** (2004) 114.
[17] L. Leonor et al., *Classical Quant. Grav.* **27** (2010) 8.
[18] M. Gromov, Neutrino 2016 conference proceedings, Conference Series (JPCS), to be published by IOP Publishing.
[19] C. Casentini, *Journal of Physics: Conference Series (TAUP 2015)* **689** (2016) 072001.
[20] R. Tomas, D. Semikoz, G.G. Raffelt, M. Kachelriess, and A.S. Dighe, *Phys. Rev.* D **68** (2003) 093013. T. Mühlbeier, H. Nunokawa, and R. Zukanovich Funchal, *Phys. Rev.* D **88** (2013) 085010.
[21] J.F. Beacom, R.N. Boyd, and A. Mezzacappa, *Phys. Rev.* D **63** (2001) 073011.
[22] T.E. Baumgarte, S.L. Shapiro, S.A. Teukolsky, *Astrophys. J.* **458** (1996) 680.
[23] K. Sumiyoshi, S. Yamada, & H. Suzuki, *Astrop. J.* **667** (2007) 382.
[24] A. Esmaili, O. L. G. Peres, P. D. Serpico, *Phys. Rev.* D **90** (2014) 033013.
[25] Private communication, Georg Raffelt, MPI Munich (2015).
[26] H. Andresen et al., (2016), arXiv:1607.05199.
[27] C. D. Ott et al., Core-Collapse Supernovae, Neutrinos, and Gravitational Waves, Proceedings of the Neutrino 2012 Conference, Kyoto, *Nucl. Phys. B (Proc. Suppl.)* **235236** (2013) 381.
[28] T.Yokozawa,*Astrophys. J.* **811**, 2015) 2.
[29] C. D. Ott et al., *Phys. Rev.* D **86**, (2012) 024026.
[30] I. Tamborra et al., *Phys. Rev.* D **90**, 045032 (2014) 4.
[31] N. Manger, bachelor thesis, Mainz 2012.
[32] J. Scargle, *Astrophys. J.* **504** (1998) 405 and J. Scargle et al., *Astrophys. J.*, **764** (2013) 167.

IOP Conf. Series: Journal of Physics: Conf. Series **1029** (2018) 012002 doi:10.1088/1742-6596/1029/1/012002

The LArIAT experiment: first measurement of the inclusive total pion cross-section in Argon

Flor de María Blaszczyk

Boston University, 590 Commonwealth Avenue, Boston 02215 USA

E-mail: fblaszcz@bu.edu

Abstract. In light of future large neutrino experiments such as DUNE, an excellent understanding of LArTPCs is required. The Liquid Argon In A Test-beam (LArIAT) experiment, located in the Fermilab Test Beam Facility, is designed to characterize the performance of LArTPCs and improve the reconstruction algorithms but also to measure the cross-sections of charged particles in Argon. The goals and experimental layout will be presented, as well as the world's first inclusive total pion interaction cross-section on Argon measured by LArIAT.

1. Introduction

Neutrinos are neutral particles, hence all their physics results rely entirely on the reconstruction of the particles coming out from their interactions with the detector. Identifying these outgoing particles requires good reconstruction capability, optimized particle identification algorithms and background control. LArIAT, the **L**iquid **Ar**gon TPC **I**n **A** **T**est-beam experiment, located in the Fermilab Test Beam Facility (Fig. 1) will allow the characterization of LArTPCs with a test beam, contributing to improve reconstruction and PID of charged particles.

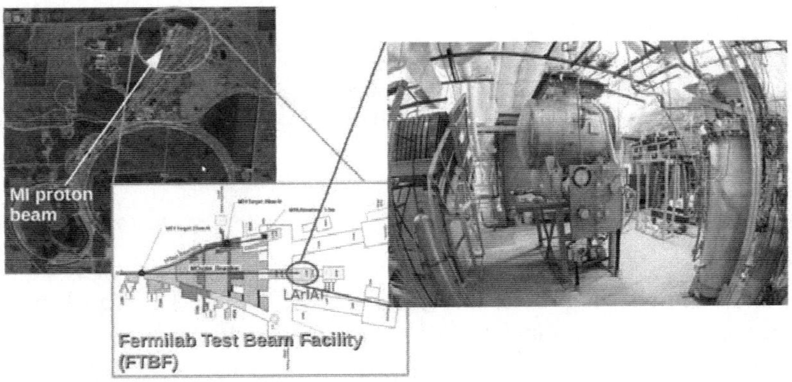

Figure 1. The Fermilab Test Beam Facility at Fermilab, Batavia IL, USA.

One of the most challenging backgrounds in neutrino electron appearance studies are neutral

IOP Conf. Series: Journal of Physics: Conf. Series **1029** (2018) 012002 doi:10.1088/1742-6596/1029/1/012002

pions, since π^0 decay photons cannot be distinguished from the electrons produced by ν_e interactions in traditional water Cherenkov and scintillator-based detectors. LArTPCs can separate these two types of particles using the first few centimeters of the electromagnetic shower: photons convert into e^+/e^- pairs thus deposit double the energy than an electron shower. LArIAT will study the electron vs photon separation power of LArTPCs.

LArIAT will also help reducing the systematic errors on neutrino related measurements by measuring hadron cross-sections on Argon. These results will contribute to validate simulation software such as GENIE[1] and GEANT4[2]. The total pion cross-section on Argon measurement with LArIAT's data will be shown.

2. The LArIAT experiment

The experiment can be divided in two parts: the test-beam with the beamline detectors, and the TPC itself.

2.1. The test-beam and beamline instrumentation

LArIAT uses a dedicated tertiary beam. The beam energy is tunable, thus can be adjusted to cover relevant energy regions for both short baseline experiments, like MicroBooNE[3], SBND and Icarus[4]; and long baseline experiments like DUNE[5].

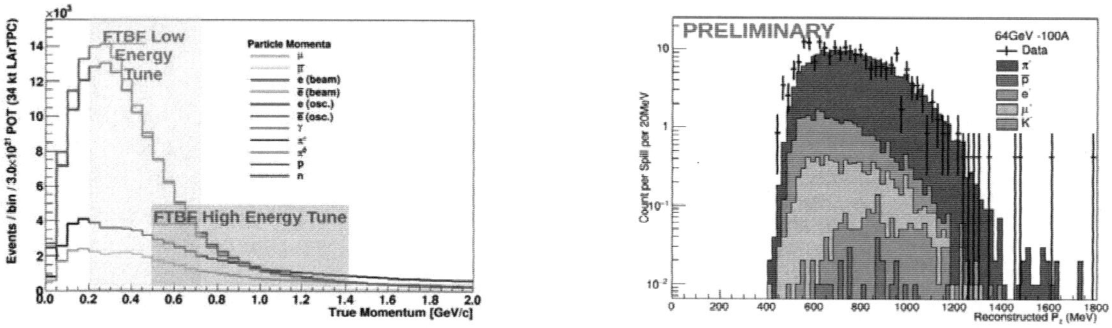

Figure 2. The left-hand plot shows a simulation of the expected outgoing particles spectra for the DUNE ν_e charged-current analysis and the energy ranges covered by the 2 LArIAT test-beam tunes. The right-hand plot shows the breakdown into particles for the LArIAT test-beam high-energy tune.

The advantage of using a test-beam is that the incoming particle type and momentum are known. For this, a series of detectors (Fig. 3) are used:

- 4 multi-wire proportional chambers (MWPC), with 2 bending magnets, for single particle momentum reconstruction, momentum and polarity selection;
- a time-of-flight system, which combined with the MWPCs information allows to do particle identification;
- 2 aerogel Cherenkov detectors with different indexes;
- a muon range stack to tag through-going particles;
- halo and punch-through vetoes.

IOP Conf. Series: Journal of Physics: Conf. Series **1029** (2018) 012002 doi:10.1088/1742-6596/1029/1/012002

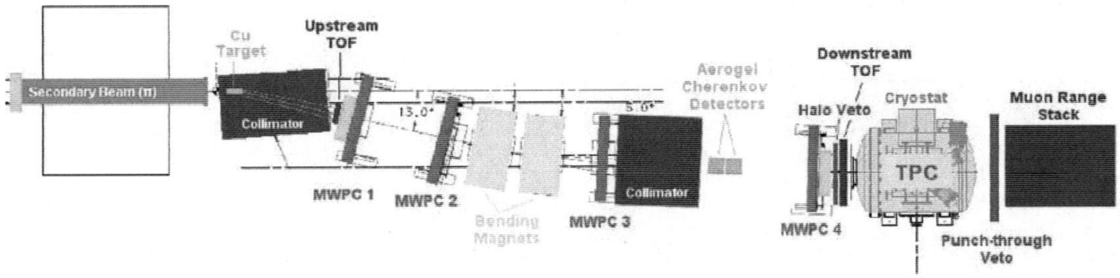

Figure 3. LArIAT beamline detectors layout.

2.2. The LArIAT TPC

LArIAT uses the refurbished ArgoNeuT[6] TPC and cryostat. The experiment characteristics are as follows:

- an active volume of 170L ($40cm \times 90cm \times 47cm$ drift), placed inside a 550L cryostat;
- a new Ti beam window on the cryostat's front flange;
- 3 wire planes (240 wires each, 4mm wire pitch), 2 of which are instrumented (induction and collection planes);
- a tunable electric field set at a nominal value of 500 V/cm;
- a light collection system made out of 2 standard cryogenic photomultipliers, 3 silicon photomultipliers, and wavelength shifting reflector foils lining the inner TPC walls;
- cold front-end electronics, which use MicroBooNE ASICs on custom motherboards;

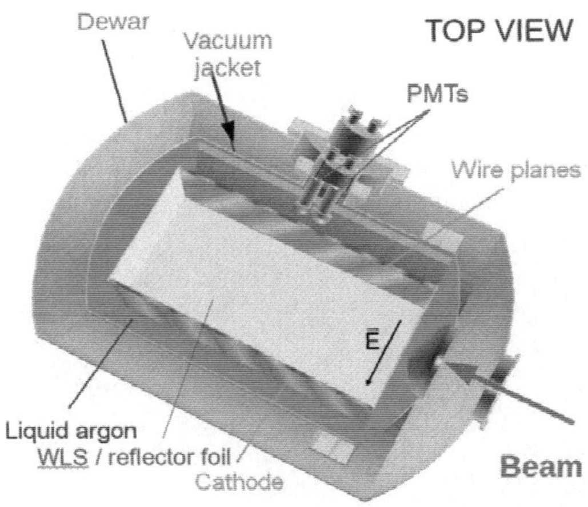

Figure 4. LArIAT schematic top view.

Figure 5. The LArIAT TPC, front-end cold electronics mounted.

The experiment has had 2 data taking periods: Run I from May to July 2015, and Run II from February to July 2016. In both cases data was taken with both beam polarities, and at both high and low energy tunes. A third run is expected to start in February 2017, focused mainly on R&D such as studying the effects of different wire pitches (3mm vs 5mm), and new light collection systems.

3. Total pion cross-section on Argon measurement

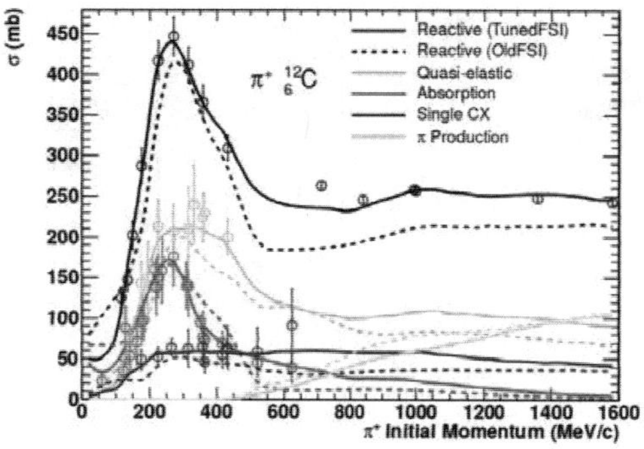

Figure 6. Pion cross-sections vs pion initial momentum. [7]

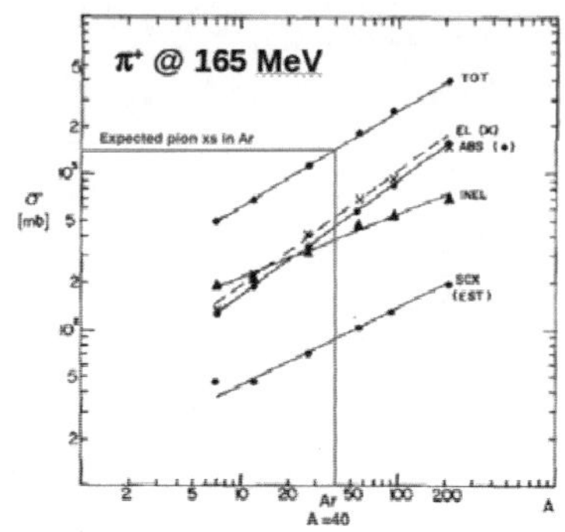

Figure 7. Total cross-section for π^+ at 165 MeV vs number of nucleons. [8]

Pions produced by neutrino interactions in experiments like DUNE are expected to be in the 100 to 500 MeV region. As shown on Fig. 6, pion cross-sections are boosted in that energy range. Because the largest systematic errors in neutrino cross-section measurements come from final state interactions (FSI) and secondary interactions, it becomes crucial to have a measurement of the different pion cross-sections. In Argon, FSI will lead to a large fraction of pions being absorbed and thus modifying outgoing particles kinematics. So far no measurement has been made on Argon and all calculations rely only on extrapolations of data taken on other nuclei (Fig. 7).

3.1. Selection

The definition of the total pion cross-section for our measurement is as follows:

$$\sigma_{tot} = \sigma_{elastic} + \sigma_{absorption} + \sigma_{charge-exchange} + \sigma_{\pi-prod} + \sigma_{ineslatic} \tag{1}$$

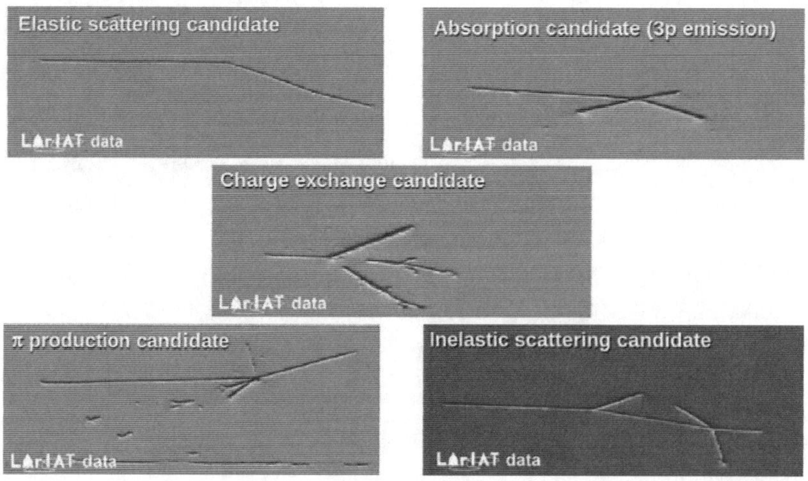

Figure 8. Pion candidates from LArIAT data.

The first step of the analysis is selecting pion-like events (Fig. 8). A pion candidate is an event where:

- the incoming particle is compatible with the pion hypothesis (combination of momentum and time-of-flight information);
- the event is not a cosmic ray;
- the reconstructed TPC track starts within the first few centimeters of the TPC;
- there is no pile-up in the event and the wire-chamber track can be uniquely matched to a reconstructed track inside the TPC;
- the event is not shower-like.

We apply these cuts to LArIAT Run I, negative polarity runs only (Tab. 1). The number of pion candidates after reduction is 2 290.

Table 1. Pion candidate selection cut flow.

	Number of events
Initial number of candidates	32 064
$\pi/\mu/e$ ID (beamline info + not cosmic)	15 448
Upstream track ($z_{init} < 2cm$)	14 330
No pile-up	9 281
MWPC/TPC track matching	2 864
Not shower-like (topology based)	2 290

From simulations, we know that the beam is composed of 48.4% π^- before the selection and of 91.04% π^- after cuts, thus the pion selection efficiency is 74.5% (Tab. 2).

Table 2. Beam composition from simulation before and after cuts.

	π^-	e^-	γ	μ^-	K^-
Before cuts	48.4%	40.9%	8.46%	2.2%	0.04%
After cuts	91.04%	3.7%	0.2%	5%	0.06%

This sample contains four types of background that have yet to be subtracted:

- pion decay, estimated at 2%;
- pion capture, ~9%;
- crossing muon contamination, ~10%;
- electron contamination, partially removed by the shower filter.

3.2. Method

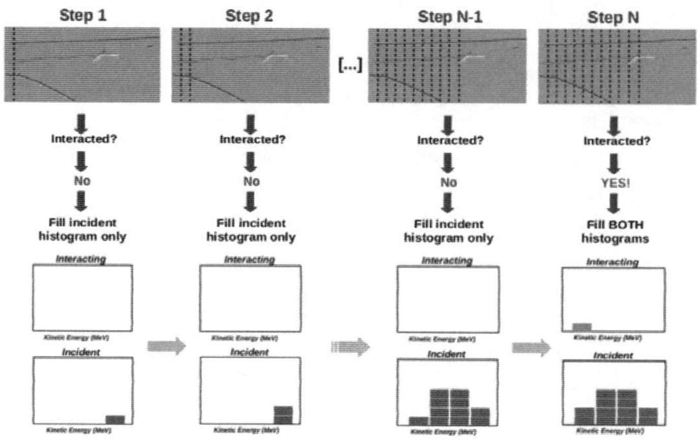

Figure 9. Thin slice method applied to a pion event in LArIAT.

To measure the total pion cross-section on Argon, we chose the thin target approach. In this case, the interaction probability for a thin target slab can be written as:

$$P_{interaction} = 1 - P_{survival} = 1 - e^{-\sigma n z} \approx 1 - (1 - \sigma n z + O(z^2)) \tag{2}$$

where σ is the cross-section per nucleon, n the medium density, and z the slab thickness.

The interaction probability can also be defined as the ratio of the number of interacting particles to the total number of incident particles:

$$P_{interaction} = \frac{N_{interacting}}{N_{incident}} \tag{3}$$

By combining Eq. 2 and Eq. 3, we obtain the expression of the cross-section as a function of energy:

$$\sigma(E) = \frac{1}{nz} \frac{N(E)_{interacting}}{N(E)_{incident}} \tag{4}$$

The Argon volume inside the TPC can be considered as a series of thin slabs, where the slab thickness is given by the TPC wire pitch: ~4 mm thick slabs in LArIAT's case. To be able to calculate the cross-section as a function of the energy, we must first calculate the pion kinetic energy at each slab. For the first slab, the initial π kinetic energy is given by:

$$KE_i = \sqrt{p^2 + m_\pi^2} - m_\pi - E_{flat} \tag{5}$$

where p is the π momentum measured by the MWPCs, m_π the pion mass, and E_{flat} the energy loss due to material upstream of the TPC. For each subsequent slab, the π kinetic energy is given by:

$$KE_{interaction} = KE_i - \sum_{i=0}^{N} \frac{dE}{dX_i} \times Pitch_i \tag{6}$$

where KE_i is given by Eq. 5, N is the number of slices, and dE/dX_i is the average pion energy loss in Argon.

In practice, the cross-section is proportional to the ratio of 2 histograms: the interacting KE histogram, which is filled with the kinetic energy of the pion when it interacts (only once per event); and the incident KE histogram, which is filled with the kinetic energy of the pion at each slice, including the slice where it interacts (Fig. 9).

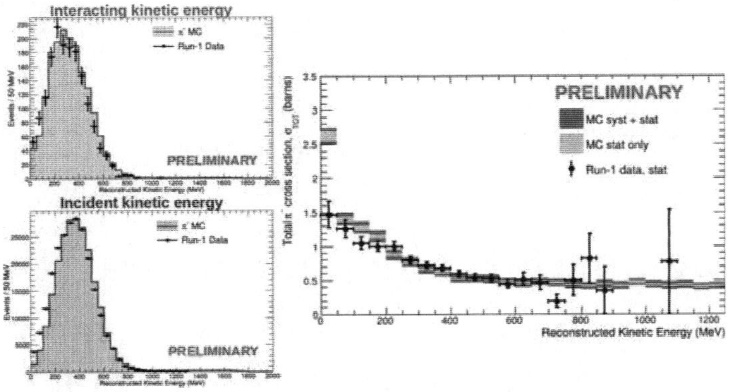

Figure 10. Top left: kinetic energy of interacting pions. Bottom left: kinetic energy of incident pions. Right: Total π^- cross-section on Argon with systematic errors.

3.3. Results

After applying the method described in the previous section, we obtain the two histograms on the left in Fig. 10. By taking the ratio of these two histograms, we extract the cross-section as a function of the pion kinetic energy shown on the right in Fig. 10. The following systematics have been included:

- dE/dX calibration error, 5%;
- uncertainty on the energy loss before entering the TPC, 3.5%;
- error due to through-going muon contamination, 3%;
- momentum uncertainty (from wire chamber reconstruction), 3%.

4. Perspectives

LArIAT runs I and II were very successful and now we are ready to start the third data taking campaign. We have observed the excellent performance of cold electronics, with a factor two improvement on the signal-to-noise ratio with respect to ArgoNeuT's warm electronics. The data gathered is being used in various analyses, including exclusive pion cross-sections, kaon cross-sections, purity measurements, and several light collection system related studies.

LArIAT has successfully measured the total pion cross-section on Argon for the first time. The analysis is still on-going, several improvements are in progress. To decrease the statistical errors, Run II data will be included as well as the positive polarity runs from Run I. To increase the efficiency of our selection, we are working on improving the MWPC track reconstruction as well as the TOF reconstruction. Backgrounds are in the process of being subtracted, and using data from the aerogel Cherenkov detectors and the muon range stack will be very helpful.

References

[1] The GENIE Neutrino Monte Carlo Generator: Physics and User Manual, arXiv:1510.05494 [hep-ph]
[2] S. Agostinelli et al., Nuclear Instruments and Methods A 506 (2003) 250-303
[3] MicroBooNE collaboration, JINST Vol. 12, P02017 (2017)
[4] C. Rubbia et al., CERN. CERN-EP/77-08 (1977)
[5] LBNF and DUNE CDR, arXiv:1601.05471 [physics.ins-det]
[6] C. Anderson et al., JINST Vol. 7, P10019 (2012)
[7] NEUT Pion Final State Interactions, arXiv:1405.3973 [nucl-ex]
[8] D. Ashery et al., Phys. Rev. C23, 2173 (1981)

IOP Conf. Series: Journal of Physics: Conf. Series **1029** (2018) 012003 doi:10.1088/1742-6596/1029/1/012003

A TPC as high performance gamma-ray telescope and polarimeter: polarisation measurement in a beam between 1.7 and 74MeV with HARPO

P. Gros[1], S. Amano[2], D. Attié[3], P. Baron[3], D. Baudin[3], D. Bernard[1], P. Bruel[1], D. Calvet[3], P. Colas[3], S. Daté[4], A. Delbart[3], M. Frotin[1,*], Y. Geerebaert[1], B. Giebels[1], D. Götz[5], S. Hashimoto[2], D. Horan[1], T. Kotaka[2], M. Louzir[1], F. Magniette[1], Y. Minamiyama[2], S. Miyamoto[2], H. Ohkuma[4], P. Poilleux[1], I. Semeniouk[1], P. Sizun[3], A. Takemoto[2], M. Yamaguchi[2], R. Yonamine[3] and S. Wang[1,]**

[1]LLR, École Polytechnique, CNRS/IN2P3, Palaiseau, France
[2]LASTI, University of Hyôgo, Japan
[3]Irfu, CEA-Saclay, France
[4]JASRI/SPring8, Japan
[5]AIM, CEA/DSM-CNRS-Université Paris Diderot, France

E-mail: `philippe.gros@llr.in2p3.fr`

Abstract. We presented in 2014 the very first data from a polarised gamma-ray beam between 1.7 and 74MeV. We now show the results of their analysis, and in particular the polarimetry measurements. With these results, we are establishing a new, high-performance way to do gamma-ray astronomy and, for the first time, polarimetry, in the e+e- pair regime.

1. Introduction

A number of groups are developing pair-conversion detector technologies alternative to the tungsten-converter / thin-sensitive-layer stacks of the COS-B / EGRET / Fermi-LAT series, to improve the single-photon angular resolution. Presently, observers are almost blind in the 1-100 MeVenergy range, mainly due to the degradation of the angular resolution of e+e- pair telescopes at low energies: to a large extent, the sensitivity-gap problem is an angular-resolution issue.

We have shown that gaseous detectors, such as TPCs (time projection chambers), can enable an improvement of up to one order of magnitude in the single-photon angular resolution (0.5°at 100 MeV) with respect to the Fermi-LAT (5°at 100 MeV), a factor of three better than what can be expected for Silicon detectors (1.0-1.5°@ 100 MeV). With such a good angular resolution, and despite a lower sensitive mass, a TPC can close the sensitivity gap at the level of 10^{-6} MeV/cm^{2}s)

*Now at GEPI, Observatoire de Paris, CNRS, Univ. Paris Diderot, Place Jules Janssen, 92190 Meudon, France

**Now at INPAC and Department of Physics and Astronomy, Shanghai Jiao Tong University, Shanghai Laboratory for Particle Physics and Cosmology, Shanghai 200240, China

between 3 and 300 MeV. In addition, the single-track angular resolution is so good that the linear polarisation fraction and angle of the incoming radiation can be measured.

We first describe the fast-gas (low pile-up), cool-gas (low-diffusion), high pressure (0.5-4 bar) HARPO detector prototype, and its test in a polarised photon beam. We quickly describe the full detector simulation developed, and its calibration using cosmic-ray measurements. Finally, we show early polarisation measurement at 11 MeV, and compare it to our simulation.

2. The HARPO Experimental setup

The HARPO (Hermetic ARgon POlarimeter) detector [1] is a demonstrator of the performance of a TPC for measuring polarised γ rays. It was designed for a validation on the ground in a photon beam. The most critical constraints related to potential space operation were taken into account, such as the reduced number of electronic channels, and long-term gas-quality preservation.

The HARPO detector is a 30 cm cubic TPC, filled with an Ar:isobutane (95:5) gas mixture at 2.1 bar. It is equipped with a drift cage providing a 220 V/cm drift field. The electrons produced by the ionisation of the gas drift along the electric field toward the readout plane at a constant velocity $v_{drift} \approx 3.3\,\mathrm{cm}/\mu\mathrm{s}$. The readout plane is equipped with two Gas Electron Multipliers (GEMs) and one Micromesh Gas Structure (Micromegas) to amplify the electrons. The amplified electrons' signal is collected by two sets of perpendicular strips (regular strips in the X-direction, and pads connected together by an underlying strip in the Y-direction. The signals from the strips are read out and digitised with a set of AFTER chips [2] and associated Front End Cards (FECs).

The HARPO TPC was set up in the NewSUBARU polarised photon beam line [3] in November 2014. The photon beam is produced by Inverse Compton Scattering (ICS) of an optical laser on a high energy (0.6-1.5 GeV) electron beam. Using lasers of various wavelengths and different beam energies (as described in more detail in [4]), we scanned 13 photon energies from 1.74 MeV to 74 MeV. In order to minimise systematic effects due to the geometry of the detector, the detector itself was rotated around the beam axis to 4 different angular positions (-45, 0, 45 and 90 °). Finally, for some configurations (in particular at low energy), data were also taken with randomly polarised photons.

3. Simulation of the HARPO detector

The cubic geometry of the detector introduces a systematic bias to the polarisation measurement which cannot be addressed analytically. It is therefore necessary to have an accurate simulation of the TPC. We developed a complete simulation to describe the response of the HARPO detector. It contains three main components:

- an event generator describing the conversion of photons in the gas [6]. It provides the energy momentum of the electron-positron pair.
- a simulation of the interaction of the electron and positron pair with the gas, based on Geant4 [5]. It provides the ionisation electrons in the gas volume.
- a custom description of the processes and geometry of the TPC [8]. It provides a signal map similar to the real data.

The first two components are already independently validated. The last was developed specifically for HARPO, and has to be validated and calibrated. The description of the TPC includes electron drift, diffusion and amplification in the gas, the readout space and time response, and the signal digitisation, including known electronics saturation effects.

This simulation was thouroughly calibrated and validated using a very tight selection of cosmic rays. Figure 1 shows an example of the calibration of the electron diffusion in the gas. This was done with a measurement of the variation of the spatial extent of the signal as a

IOP Conf. Series: Journal of Physics: Conf. Series **1029** (2018) 012003 doi:10.1088/1742-6596/1029/1/012003

function of drift distance. We see a perfect agreement between data and simulation. All the simulation parameters were similarly calibrated or validated against data.

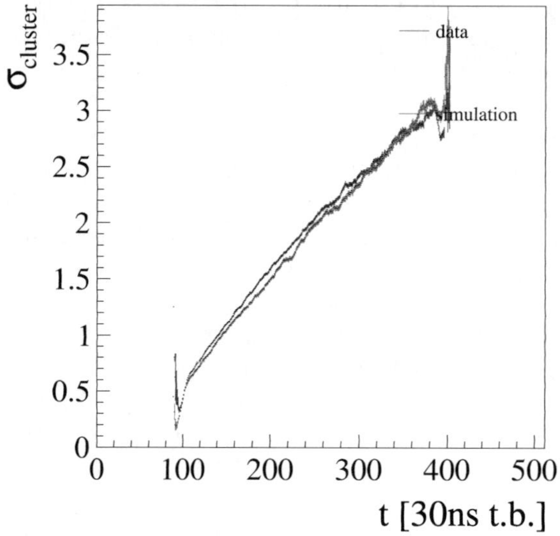

Figure 1. Signal size (RMS) as a function of the drift distance for data (red) and simulation (blue), for a sample of cosmic rays. There is excellent agreement between the data and simulation.

4. Results on polarimetry

A specific reconstruction algorithm was developed to reconstruct vertices in the TPC [8]. This allows us to extract the azimuthal angle ϕ_{+-} for the conversion [7],

$$\phi_{+-} = \frac{\phi_+ + \phi_-}{2}, \tag{1}$$

where $\phi_\pm$ are the azimuthal angles (relative to the beam direction) of the electron and positron. ϕ_{+-} is expected to follow a distribution

$$1 + A\cos 2(\phi_{+-} - \phi_0)$$

, where A is the polarisation asymmetry, and ϕ_0 is the polarisation direction.

We cancel the systematic effects by taking the ratio of the polarised and unpolarised distributions of the azimuthal angle ϕ_{+-}. Figure 2 shows the resulting distributions for data and simulation. We can also correct the systematics in the data using the simulation. The result is also shown in Fig. 2.

5. Conclusions and outlook

The HARPO detector was completed and used successfully in a photon beam in 2014. A full simulation of the detector was developed to describe it, and it was validated and calibrated against cosmic-ray data. We measure the polarisation asymmetry of a 11.8 MeV photon beam with high precision, and with good agreement with simulation.

There is still ongoing efforts to improve these results, and extend them to the full energy range (1.74 to 74 MeV) of the available beam data. The results from the beam data are extremely encouraging and studies are under way to adapt this technology for use as a space telescope.

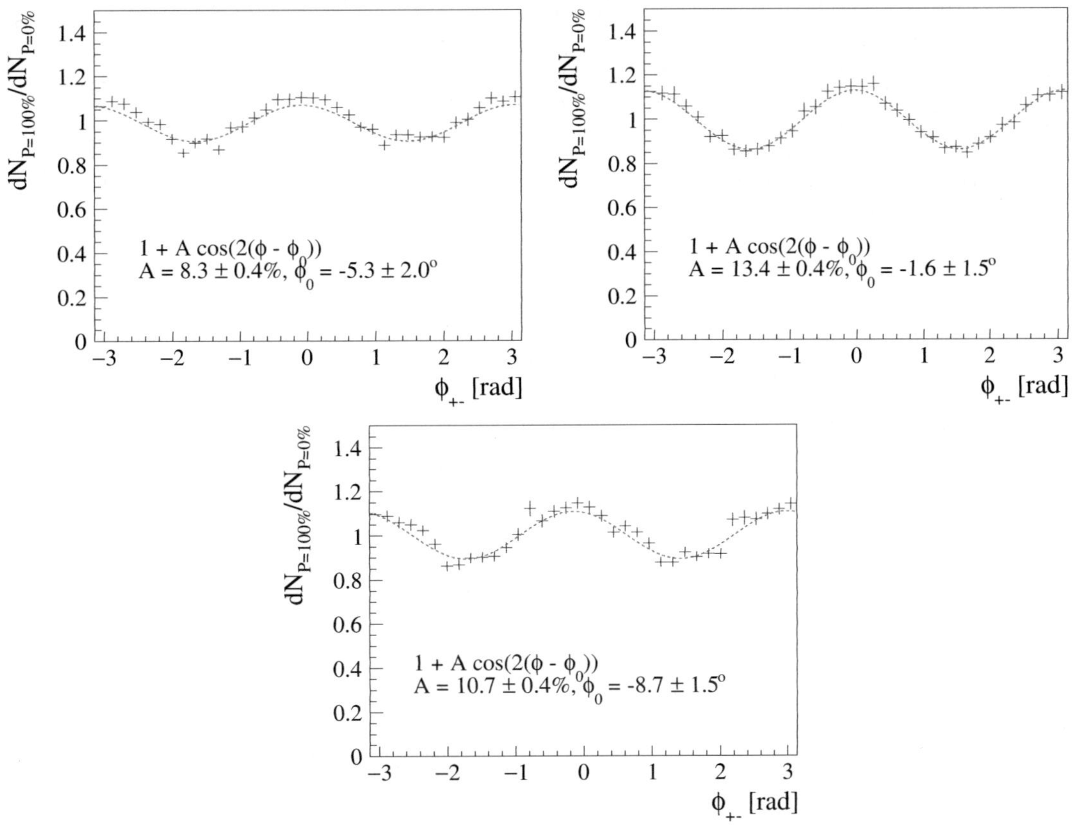

Figure 2. Polarisation modulation for a beam of 11.8 MeV photons. The plot is obtained by dividing the distribution of the azimuthal angle ϕ_{+-} for polarised photons, by the distribution for unpolarised photons. Most of the systematic errors are cancelled, only statistical errors are shown. On the top left, both distributions are for real data. On the top right, both distributions are simulated. On the bottom, the polarised distribution comes from real beam data, and the unpolarised one comes from simulation.

Acknowledgments
This work is funded by the French National Research Agency (ANR-13-BS05-0002) and was performed by using NewSUBARU-GACKO (Gamma Collaboration Hutch of Konan University).

References
[1] P. Gros *et al.*, "HARPO - TPC for High Energy Astrophysics and Polarimetry from the MeV to the GeV," *Proceedings of Science*, vol. TIPP2014, p. 133, 2014.
[2] D. Calvet, "A Versatile Readout System for Small to Medium Scale Gaseous and Silicon Detectors," *IEEE Trans. Nucl. Sci.*, vol. 61, no. 1, pp. 675–682, 2014.
[3] K. Horikawa, S. Miyamoto, S. Amano, and T. Mochizuki, "Measurements for the energy and flux of laser Compton scattering γ-ray photons generated in an electron storage ring: NewSUBARU," *Nuclear Instruments and Methods in Physics Research Section A: Accelerators, Spectrometers, Detectors and Associated Equipment*, vol. 618, no. 1-3, pp. 209 – 215, 2010.
[4] S. Wang *et al.*, "HARPO: a TPC concept for γ-ray polarimetry with high angular resolution in the MeV-GeV range," *J. Phys. Conf. Ser.*, vol. 650, p. 012016, 2015.
[5] J. Allison *et al.*, "Geant4 developments and applications," *IEEE Transactions on Nuclear Science*, vol. 53, pp. 270–278, Feb 2006.

[6] P. Gros and D. Bernard, Astropart. Phys. **88** (2017) 60 doi:10.1016/j.astropartphys.2017.01.002 [arXiv:1612.06239 [astro-ph.IM]].

[7] "γ-ray polarimetry with conversions to e^+e^- pairs: polarisation asymmetry and the way to measure it", P. Gros and D. Bernard, accepted for publication by Astroparticle Physics, doi 10.1016/j.astropartphys.2016.12.006, arXiv:1611.05179 [astro-ph.IM].

[8] "Measurement of polarisation asymmetry for gamma rays between 1.7 to 74 MeV with the HARPO TPC", P. Gros *et al.*, SPIE2016, 9905-95.

IOP Conf. Series: Journal of Physics: Conf. Series **1029** (2018) 012004 doi:10.1088/1742-6596/1029/1/012004

On the possibility of positive-ion detection in gaseous TPCs and its potential use for neutrinoless double beta decay searches in ^{136}Xe

Lior Arazi

Nuclear Engineering Unit, Faculty of Engineering Sciences, Ben Gurion University of the Negev, P.O.B. 653 Beer-Sheva 8410501, Israel

E-mail: larazi@bgu.ac.il

Abstract. The neutralization of slow positive ions on solid surfaces can lead to the emission of secondary electrons in Auger-type processes. We discuss the possibility of harnessing such mechanisms to the detection of positive ions in gaseous TPCs. Applied to high pressure xenon, the proposed idea may enable reconstructing with high accuracy the topology of candidate neutrinoless double beta decay events of ^{136}Xe without sacrificing the energy resolution of pure Xe gas. Candidate secondary electron emitters are discussed, as well as the expected efficiencies, challenges and potential pitfalls.

1. Introduction

Time projection chambers (TPCs) are a central tool in experimental high energy physics, nuclear physics and astroparticle physics, providing detailed three dimensional images of ionization tracks produced by charged particles passing through their sensitive volume. In the basic TPC scheme, ionization electrons liberated along the track are drifted towards a position-sensitive readout plane, where the track image is reconstructed. While in most applications this principle is sufficient, for some experiments traverse and longitudinal diffusion of the drifting electrons presents a serious limitation. This is especially true in cases where on the one hand the physics case requires resolving spatial features on the mm or sub-mm scale, and on the other hand the TPC dimensions must be large (meter-scale) to contain the full track or to allow for sufficient sensitivity, as in rare-event searches.

One way to suppress electron diffusion in gaseous TPCs is to apply a strong magnetic field parallel to the drift direction [1], but this is not always a practical or economic solution. Alternatively, one may use gas mixtures with small electron diffusion coefficients, but with a potential compromise of other properties of the detector, such as energy resolution [2]. A different approach, which may be applied if the event rate is sufficiently low, is to rely on *ions* to store the topology of the original track and transport it slowly and with minimal diffusion to an ion-sensitive readout plane at the endcap of the TPC. The Negative Ion TPC (NITPC) concept, developed by Martoff *et al* [3], implements this idea by using electronegative molecules to rapidly capture the electrons generated along the ionization track. The resulting negative ions are then drifted towards a readout plane where they are stripped of their extra electrons, which subsequently undergo avalanche multiplication in an appropriate amplification structure. The concept of a low-pressure NITPC is pioneered by the DRIFT program for directional dark matter detection [4], where few-mm long ionization tracks of low-energy ($\lesssim$100 keV) nuclear recoils must be reconstructed with sub-mm accuracy to establish head-tail asymmetry. The collaboration's current detector, DRIFT-IId, uses a gas mixture of 30:10:1 Torr Cs_2: CF_4: O_2, where fluorine atoms act as the target for WIMP scattering and CS_2 is the electronegative gas; avalanche multiplication is done on wires. An NITPC scheme is also under development for x-ray polarimetry of astronomical sources [5], using nitromethane (CH_3NO_2) as the electronegative gas and Gas Electron Multiplier (GEM) [6] structures for charge amplification.

The possibility of applying the NITPC concept to ton-scale searches for neutrinoless double beta decay ($0\nu\beta\beta$) in ^{136}Xe was suggested by D. Nygren in [7]. The TPC, in this case, would consist of high pressure (~10-20 bar) Xe gas enriched in ^{136}Xe with a small admixture of electronegative gas. The idea is to capture the track electrons by the electronegative molecules, drift the negative ions to the anode plane and there – since they drift very slowly - count them one by one, providing near-intrinsic energy resolution and accurate spatial information. A main challenge in realizing this concept is the identification of a suitable electronegative molecule, with electron affinity high enough to keep the extra electron bound during the drift, and low enough to release it in the amplification region at the anode plane. A second possible complication is the loss of the primary scintillation (and hence t_0) signal due to UV absorption either by the electronegative admixture itself or by other additive molecules which may be required for gain stabilization. As suggested in [7], t_0 may possibly be recovered using 'echo' signals generated by positive Xe ions on the cathode through the release of secondary electrons, which would be quickly captured and drifted to the anode by the electronegative molecules.

In this contribution we look (for now – theoretically) into the possibility of *positive* ion detection in TPCs as a complementary approach to suppress electron diffusion. The idea of extracting spatial information about the track from its positive ions is not new, but has so far been applied only in very low-pressure scenarios. Positive ion counting, relying on differential pumping of low pressure gas (< 1 Torr) exposed to ionizing radiation, was first applied in the early 1970s [8, 9], and later in the 1990s and 2000s [10-13], to measure the size distribution of ionization clusters. The basic idea was to simulate nanometer-scale ionization damage to the DNA by recording the structure of mm-scale clusters in a tissue-equivalent gas. The ions were electrostatically extracted from the low-pressure gas volume to vacuum, where they were detected by a micro-channel or micro-sphere plate. More recently, an idea for position-sensitive detection of positive ions by impact ionization of gas molecules at low pressure (few Torr) using a Thick GEM-like structure [14] was suggested in [15]. Here the main difficulty is to accelerate the ions to sufficient energy between collisions, necessitating operation at low pressure and the use of very thick structures [16].

In what follows we discuss the possibility of positive ion detection in atmospheric and high-pressure gaseous TPCs, using potential emission of secondary electrons from surfaces by the impinging ions. The emitted electrons can be used to generate localized light or charge signals, recorded by position-sensitive readout. We note that experimental work on *internal* potential emission (where valence electrons are excited to the conduction band) by positive ions was recently reported on in [17], and that an idea of using external potential emission for positive ion detection in low-pressure TPCs for nanodosimetry studies was recently raised by V. Dangendorf independently of this proposal [18]. We begin with an overview of potential emission processes, followed by general considerations for employing them in gaseous TPCs and a discussion of several candidate materials. We then turn to look into the possible application of the idea to the detection of Xe ions in high-pressure TPCs for $0\nu\beta\beta$ searches in ^{136}Xe, highlighting the particular requirements and challenges in this case.

2. Ion-induced potential emission of secondary electrons from surfaces

The neutralization of a slow positive ion on a clean solid surface occurs primarily by means of two basic mechanisms: resonance neutralization (RN) and Auger neutralization (AN) [19-24]. If the ionization energy of the ion is sufficiently large, AN can lead to the emission of a secondary electron from the surface, in addition to the one that neutralizes the ion. A secondary electron may also be emitted in the case of RN if the resonant transition is to an excited atomic state and is followed by Auger de-excitation (AD) as explained below. Both AN and RN (followed by AD) can take place on metals, semiconductors or insulators (although for the latter the picture may be more complicated [22]). Both can occur for vanishingly small ion kinetic energies - it is the potential energy recovered during neutralization which serves to eject the electrons from the surface; as such, they are referred to as *potential emission* processes, in contrast to kinetic emission processes which are driven by the ion kinetic energy (typically at the keV scale).

RN is a one-electron process which is allowed if the ground state or one of the excited states of the approaching ion is in resonance with one of the filled continuum energy levels in the solid (i.e., if they lie opposite to a filled band). RN to the ion ground state does not generally lead to the emission of a secondary electron from the surface (Figure 1A) [21]. However, when tunneling is to an excited atomic state, de-excitation of the atom can be accompanied by the simultaneous emission of a second electron from the surface – provided that its energy is larger than the vacuum level and that its velocity is directed outward (AD, Figure 1B). A necessary condition for external electron emission by AD is that the energy of the excited state E_x' relative to the atomic ground state is larger than a minimal energy barrier φ, which for metals is the work function and for insulators is the sum of the band gap energy E_g and electron affinity χ. The maximum kinetic energy of the emitted electron, relative to the vacuum level, is: $E_k^{max} = E_x' - \varphi$.

AN is a two-electron process in which one electron tunnels out of the surface to the ion ground state, and at the same time energy and momentum are transferred to a second electron on the surface and excite it to a higher level. Unlike RN, AN can also occur if the ion ground state lies below the bottom of the filled band (Figure 1C). As in AD, this electron may then leave the surface, depending on the energy it gained and the direction of its motion. A necessary condition for external emission by AN is that the ionization energy of the ion close to the surface (i.e., the energy difference between the ground state and vacuum level) is larger than twice the abovementioned energy barrier: $E_i' > 2\varphi$. The maximum kinetic energy of the emitted electron in this case is: $E_k^{max} = E_i' - 2\varphi$ [1]. AN is possible only if the energy transferred to the surface electron can take it to an allowed state (or above the vacuum level); transitions to either filled energy levels or into the band gap (in the case of insulators) are forbidden [22]. Moreover, even for allowed transitions, external emission of the secondary electron in both AN and AD is not guaranteed, as it may remain inside the solid; this can be referred to as *internal electron emission*. Lastly, if, in addition to AN, RN to the ground state is also possible, the probability for electron emission decreases. Note that de-excitation by photon emission is negligible in AN and AD because the radiative lifetimes are larger by ~6 orders of magnitude than those of electron transitions [21].

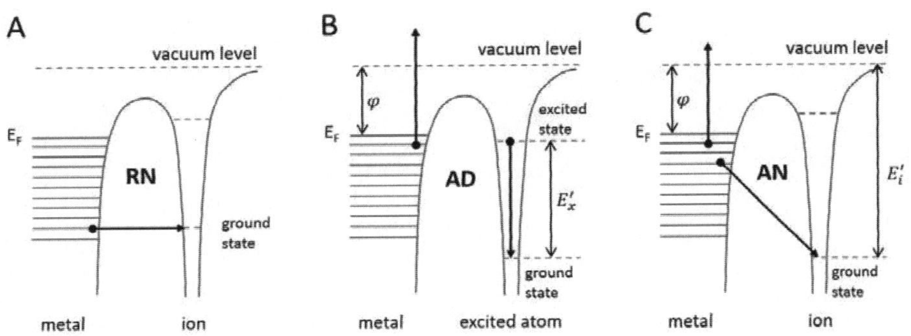

[1] The energy levels of the ion are shifted upward close to surface (in the case of a metal) due to the interaction between the ion and its image charge; in other words, the ionization energy near the surface is somewhat lower than its free-space value. (This is the reason for writing E_i' and E_x' rather than E_i and E_x.) However, this effect, which increases with the ion kinetic energy, seems to be of minor importance for slow ions [19] and may behave differently (even with an opposite sign) in insulators [22].

IOP Conf. Series: Journal of Physics: Conf. Series **1029** (2018) 012004 doi:10.1088/1742-6596/1029/1/012004

Figure 1: Transition diagrams for a metal target and an ion or excited (metastable) atom close to its surface. E_F is the Fermi energy and φ is the work function. (A) Resonance neutralization (RN) to the empty ground state of the ion, with no further electron emissions. (B) Auger de-excitation (AD) of an excited atom close to the surface. The de-excitation energy is simultaneously taken by a surface electron which, in this case, is excited above the vacuum level. (C) Auger neutralization (AN): an electron tunnels out of the surface to the empty ground state of the ion. The energy lost by this electron is simultaneously taken by a second electron on the surface, exciting it above the vacuum level.

Ion-induced potential emission of electrons from solid surfaces was extensively studied by H. D. Hagstrum in the 1950s and 1960s, both experimentally and theoretically [19, 24-30]. He focused on singly-charged atomic ions of noble gases (He^+, Ne^+, Ar^+, Kr^+, Xe^+) impinging on atomically clean metals (initially W and Mo) and on semiconductors (Si and Ge). His experiments were done in ultra-high vacuum (UHV) and the samples were atomically cleaned by flash-heating to very high temperatures (~2200 K for the refractory metals) immediately before ion bombardment. His data cover ion kinetic energies in the range 10-1000 eV. These studies, which served as the basis for all subsequent work in the field, lead Hagstrum to the advent of Ion Neutralization Spectroscopy (INS) as a sensitive probe for investigating the state densities at the solid surface by measuring the kinetic energy spectra of secondary electrons emitted by AN of slow noble gas ions [30].

Hagstrum showed that the yield γ_i of ion-induced secondary electron emission (IISEE) (i.e., the average number of electrons emitted per ion) on atomically clean targets does not change appreciably with the ion kinetic energy over the range 10-1000 eV, supporting the interpretation that the emission is governed in this range by potential energy effects [25, 26, 28]. He demonstrated that γ_i grows strongly with the difference $E_i - 2\varphi$ - particularly for Xe^+, Kr^+ and Ar^+ - starting from ~1-2% for Xe^+ and reaching ~30% for He^+ on both Mo ($\varphi = 4.3$ eV) and W ($\varphi = 4.5$ eV). His data for Mo and W targets and 10 eV ions are shown in Figure 2, which also lists the free-space ionization energy of the ions. Later works by Hagstrum and other authors on additional atomically clean metal surfaces (Cu, Ni, Au, Pt, Ta) showed electron yields of the same order of magnitude and of similar qualitative dependence on the ionization energy, but with generally smaller values than those for Mo and W [30, 31].

Figure 2: Adapted from [26] – Hagstrum's measured ion-induced secondary electron yields on atomically clean Mo and W targets, as a function of the difference between the ionization energy of the ion in free space and twice the work function of the surface. Data shown for 10 eV ions.

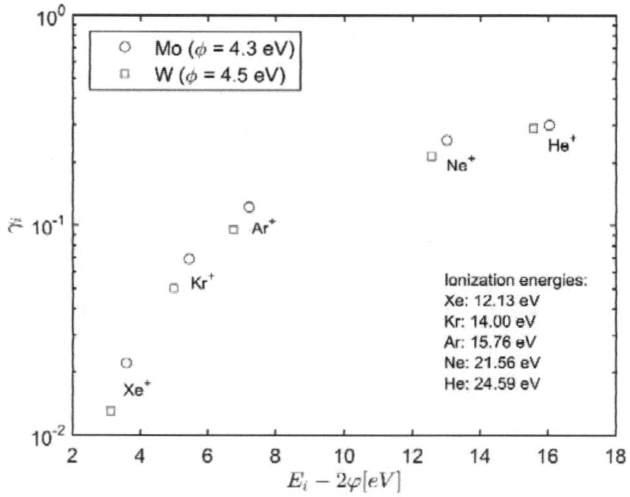

Hagstrum's measured IISEE yields for Ge and Si (for the same noble gas ions) were substantially smaller than those for Mo and W, by a factor which strongly increased with decreasing ionization energy (or with $E_i - 2\varphi$): ~1.5 for He^+, ~2 for Ne^+, ~4 for Ar^+, ~10 for Kr^+ and ~40 for Xe^+ [28]. The smaller yields are consistent with the somewhat larger energy barriers for electron emission of Si and Ge. The band gap energies for Si and Ge are 1.14 eV and 0.67 eV, respectively, with similar electron affinities: 4.05 eV for Si and 4.0 eV for Ge. The corresponding energy barriers are therefore 5.2 eV for Si and 4.7 eV for Ge. In addition, the difference in yields may also be partly due to the large width of the semiconductors valence band, which results in increasing competition between AN and RN to the ground state for ions with lower ionization energy (whereas for the refractory metals AN was the only available neutralization channel) [29].

A central observation by Hagstrum was that the presence of an adsorbed foreign monolayer on the target surface substantially reduces electron emission, by a factor which, again, increases with decreasing ionization energy [27]. The effect was studied by controllably exposing an atomically clean W target to N_2, H_2 and CO during bombardment by singly-charged noble gas ions. In all cases the rate of electron emission decreased steadily during the formation of the adsorbed monolayer and leveled off at a final steady-state value. For example, for a W target with N_2 monolayer and 10 eV ions, the suppression factor was 1.6 for He^+ and Ne^+, 2.1 for Ar^+, 2.9 for Kr^+ and 6.5 for Xe^+. Electron emission could be recovered to its original value by re-heating the W sample to 2000 K in the presence of N_2 and CO (but not H_2). The suppression of IISEE by an adsorbed monolayer of foreign material was repeatedly confirmed in later works with precisely contaminated surfaces [21, 23, 31]. The effect may be partly attributed to an increase in the effective work function of the surface, as electrons are more tightly bound to the contaminating molecules, such that the condition $E_i' > 2\varphi$ is not satisfied [23].

As discussed above, Hagstrum focused on IISEE by atomic ions of noble gases. Subsequent works showed that potential emission was also possible for *molecular* ions. Vance *et al* [32, 33] studied this for atomically clean Mo, using H_2^+ ($E_i = 15.43$ eV), N_2^+ (15.58 eV), O_2^+ (12.07 eV) and NO^+ (9.26 eV) with kinetic energies extending down to 30 eV. In this case, γ_i values were considerably smaller than those for ions of noble gas atoms of similar ionization energy (3.3% for H_2^+, 2.3% for N_2^+, and 0.7% for O_2^+ at 30 eV, ~0.1% for NO^+ at 50 eV) and displayed stronger, monotonically increasing, dependence on the ion kinetic energy. This different behavior is not fully understood and may be related to the denser energy structure of the molecules compared to single atoms [20].

Compared to metals and semiconductors, IISEE experiments on insulating targets are more difficult to perform because of charging-up effects during ion bombardment. Furthermore, additional energy dissipation channels are possible, complicating the interpretation of such experiments. Nevertheless, it appears that the basic processes of RN, AD and AN also play a dominant role in the case of insulators, with electron transitions occurring from the valence band (as in semiconductors) [22]. Substantial work on IISEE from insulating films was done during the 1990s and 2000s in the context of the development of AC plasma display panels (PDP). The aim was to optimize the discharge characteristics of these devices (which were operated with various combinations noble gases, mostly Ne/Xe mixtures), by coating the inner surface of the glass-covered cathode with a thin and durable dielectric film, exhibiting a high yield of IISEE. The primary focus was on MgO, which combines high IISEE yields with a high degree of sputter resistance, and was therefore the material of choice in most PDP cells. IISEE of slow noble-gas ions (with a typical energy of a few tens of eV) on MgO was measured by several groups, showing a large spread in γ_i, as summarized in [34]: He^+ 13-35%, Ne^+ 6-50%, Ar^+ 0.9-6%, Kr^+ 0.9-1.5%, Xe^+ 0.2-4%. This spread is attributed to complications in the experimental methodologies employed, the quality of the MgO surface and the studied crystal plane (with MgO(111) giving the highest yields) [22]. Experiments were done either using a beam of slow ions in UHV setups (as was done by Hagstrum on metals and semiconductors), or in discharge experiments in which γ_i was inferred from measuring the breakdown voltage in noble gases at sub-atmospheric pressures. The former line of experiments is complicated by charging-up effects, and the latter by the difficulty in separating the contribution to secondary electron emission from ions (of different species), photons and metastable atoms. A particularly careful study was carried out by Matulevich *et al* [35,36] on 1-5 nm thick MgO(100) on Mo(100) using a beam of 40 eV singly-charged noble gas ions with very low current to avoid charging-up effects. The measured yields in this case were: He^+ 55 $\pm$ 5%, Ne^+ 41 $\pm$ 4%, Ar^+ 10 $\pm$ 1% and Xe^+ 3 $\pm$ 2%. Based on theoretical model simulations the authors concluded that these values are too high to be explained by AN alone; this is particularly evident for Xe, where the condition $E_i' > 2\varphi = 2(E_g + \chi)$ is not met (at the MgO surface $E_g = 6.8$ eV and $\chi = 0.85$ eV [34]. They suggested that electron emission is enhanced by a process in which the remaining hole is transported across the thin MgO layer towards the Mo substrate, where its neutralization by an Auger transition results in the emission of another electron that may leave the surface.

Several groups investigated the suitability of other dielectric layers as secondary electron emitters in the context of PDP development. In particular Bachmann *et al* [37, 38] inferred secondary electron yields for hydrogen-terminated CVD diamond from discharge experiments with Ne, Ar and Xe, finding values of ~50%, ~20% and ~2%, respectively. The authors discuss the possibility that other processes (such as electron emission by UV photons) contribute to the measured yield, but conclude that it is dominated by ion-induced emission. The study compared between hydrogen-terminated CVD diamond (with negative electron affinity NEA) and oxygen-terminated CVD diamond (with positive electron affinity), showing much higher yields for the former. The NEA value reported is -0.8 eV [37], giving (for a band gap of 5.5 eV) an energy barrier $\varphi = 4.7$ eV.

Lastly, very high γ_i values were inferred from measured gain curves by Lyashenko *et al* [39] for bi- and mono-alkali photocathodes (K–Cs–Sb, Na–K–Sb, and Cs–Sb) in 700 Torr Ar/5%CH_4. In this case *thermal* ions (presumably CH_4^+) impinging on the photocathodes (with extremely low work functions ranging from 0.5 to 1.1 eV) resulted in γ_i values between 47% and 49% (with large uncertainties).

3. General considerations for employing ion-induced potential emission in TPCs

Since potential emission can occur for thermal ions (as demonstrated in [39]), it can conceivably be used to detect positive ions in gaseous TPCs operated at atmospheric or high pressure, thus opening new experimental possibilities. The basic process would be the release of secondary electrons by ions reaching the cathode plane, with subsequent detection of the emitted electrons through charge or light signals. For this to become a reality, however, several important considerations should be taken into account.

Surface purity: As discussed above, the presence of an adsorbed monolayer of foreign molecules on the surface of the target can drastically reduce the IISEE yield. It therefore appears that IISEE target materials could only be used in ultra-pure gases which either do not stick to the target surface or can be easily removed. Even then, rapid contamination of the surface seems unavoidable. Consider, for example, a TPC operated with a pure noble gas at 1 bar, containing impurities at the ppb level. The partial pressure of these impurities will be ~10^{-6} Torr which, in vacuum, would result in the formation of a monolayer within several seconds. It seems that in order to keep the surface clean, periodic heating to high temperatures will be mandatory. This could be done, for example, by depositing the target material on thin resistive wires or strips (on an insulating substrate) and apply flash ohmic heating. The wires themselves, of course, could serve as the target material (e.g., Mo or W wires). In addition to periodic flash-heating, the wires could be heated on demand following a trigger (for example a prompt scintillation-based t_0, and/or electron-based signal coming from the anode), making use of the slow drift velocity of the ions to have the target surface clean at the expected ion time of arrival. The required temperatures and periodicity of the heating cycles will vary between different types of IISEE target materials, and may expected to be lower for chemically-inert surfaces, such as MgO or H-terminated diamond. When designing such heating cycles, care should be taken to minimize convective flows inside the TPC that may interfere with the ion drift.

Ion species reaching the cathode: If the TPC contains a mixture of several gases, due to collisional charge exchange the positive ions reaching the cathode will be those of the species that has the lowest ionization energy. In pure noble-gas TPCs ions formed along the ionization track quickly convert through three-body collisions to molecular (dimer) ions through the reaction: $X^+ + 2X \rightarrow X_2^+ + X$ [40], which have slightly lower ionization energy compared to the atomic ions [34]. Although, to the best of our knowledge, no experimental data is available on the IISEE yields for noble gas dimer ions, they are likely lower than those of the corresponding atomic ions (as was observed for other molecular ions). The picture may be further complicated by the formation of clusters of neutral atoms around the drifting ions [41-45]. This can lead to dispersion in drift velocities if there is more than one dominant cluster type. The importance of this effect may vary greatly between different gases and operating pressures and should be studied in detail for any particular application.

Electron emission energy barrier: As discussed above, the IISEE yield strongly increases with the difference between the ionization energy and twice the energy barrier for electron emission φ. For most singly-charged ions (except He^+ and Ne^+) target materials with low value of φ, preferably ~3-4 eV or below, will be necessary for having an appreciable detection efficiency.

Ion detection efficiency and potential applicability: The maximum reported values of γ_i are ~50%. For many ion/target combinations yields on the few-percent level should be expected. Backscattering of the emitted electrons by elastic collisions within the gas will further reduce the effective yield in most cases. This apparently limits the potential scope of applicability to experiments in which track reconstruction can be carried out using a fraction of the primary charge, or where useful information can be obtained from the ions time of arrival. Note that the information extracted from the track's positive ions should, in most circumstances, come in addition to that gathered from its electrons.

Electron extraction into gas: As noted above backscattering by elastic collisions with the gas molecules will reduce the effective yield, which can be expressed as $\gamma_{eff} = \gamma_i \cdot \varepsilon_{ext}(gas; E/p, \bar{E}_k)$ [39]; ε_{ext} is the extraction efficiency into gas, i.e., the probability that the emitted electron is not backscattered to the cathode. The extraction efficiency depends on the gas, the reduced field E/p on the cathode surface and the average kinetic energy of the emitted electrons, $\bar{E}_k$. At a given reduced field, it is generally low for pure noble gases, for which elastic collisions are dominant at low kinetic energies, and higher for molecular gases where inelastic scattering is important. The extraction efficiency increases monotonically with E/p but generally at a monotonically decreasing rate. For some gases (e.g., Ar, Kr, Xe) it *decreases* with the kinetic energy of the emitted electron because the elastic scattering cross section increases with energy (for kinetic energies above the Ramsaeur–Townsend minimum of the elastic cross-section) [46]. In this case a lower work function will not necessarily lead to a higher effective IISEE yield because the kinetic energies of the emitted electrons will be higher, resulting in lower extraction efficiency.

Electron field emission: Low work function materials may also show a high probability for non-thermal electron field emission (EFE). Since ions arrive one by one and emit at most one electron per neutralization, individual EFE events will look the same as IISEE events.

Readout scheme: The readout system employed must be able to detect individual electrons emitted by the impinging ions. If the detection is based on avalanche multiplication, the required charge gain should be high (~10^4-10^5) depending on the readout electronic noise. This will be difficult to achieve in TPCs with a pure noble gas, requiring the addition of a quencher to avoid discharges. As discussed below for the case of high-pressure Xe, single-electron sensitivity appears to be possible by light detection, in a structure combining small charge gain and electroluminescence (EL). This can be generalized to other gases with high EL yields.

4. Candidate IISEE target materials for gaseous TPCs

The ideal IISEE material should combine several properties: (1) low energy barrier for electron emission (~3-4 eV or below); (2) low probability of electron field emission (EFE); (3) high degree of chemical stability, or (4) suitability for *in situ* surface cleaning by flash heating; (5) suitability for thin wire/strip geometries.

Mo and W appear to be the most suitable candidate metallic IISEE materials. For both metals (when their surface is atomically clean) AN is considered to be the only available neutralization channel, with no competition from RN to the ion ground state; this results in IISEE yields ranging from 2% for Xe^+ on Mo, through ~10% for Ar^+ to ~30% for Ne^+ and He^+ for both metals - in spite of their somewhat high work function (~4.3-4.5 eV for the proper crystal planes). Both Mo and W (especially W) are suitable for forming wire planes, and both are suitable for *in situ* flash ohmic heating to very high temperatures – which is likely unavoidable as their surface is not chemically inert. For ions of low ionization energy Mo appears to be preferable over W. There is no known EFE problem with either Mo or W.

MgO may be a promising option – especially if applied as an ultra-thin layer on Mo. As noted above, 1-5 nm thick MgO on Mo displayed very high IISEE yields: 55% for He^+, 41% for Ne^+, 10% for Ar^+ and 3% for Xe^+, in spite of the fact that for Xe the AN condition $E_i' > 2\varphi$ is not satisfied (assuming the nominal work function of MgO). While the investigators suggest that this may reflect an enhancement of AN by an additional mechanism [35, 36], it may also originate from an effective reduction in the Mo work function. A recent study demonstrated a work function of 3.2 eV for an 8-atomic layer thick film of MgO on Mo(001) [47], while density functional calculations suggest that applying 1-3 atomic layers of MgO on Mo(001) can lead to a work function of 2.05 eV [48]. In addition to the possibility of a very low work function, MgO surfaces are essentially chemically inert – making it an apparently ideal IISEE candidate; if some *in situ* heating is required, it can be done easily by ohmic heating of the Mo substrate. It remains to be demonstrated that ultra-thin MgO films on Mo wires or thin strips can have similar properties with no EFE issues. As long as the use of MgO is limited to low-rate (e.g., rare-event) experiments, charging up effects are expected to be minimal.

Diamond layers with negative electron affinity (NEA) are an additional interesting option. As discussed above, high IISEE yields were inferred from discharge studies for H-terminated CVD diamond with NEA of -0.8 eV (i.e., $\varphi = 4.7$ eV), namely ~50% for Ne, ~20% for Ar and ~2% for Xe [37, 38]. Since in this case RN to the ion ground state is possible (and therefore competes with AN without electron emission), these high values may have been affected, to some extent, by additional mechanisms such as photoemission, which can be quite large for the relevant VUV wavelengths [49]. A recent study [50] demonstrated a very large NEA of -2.0 eV for Mg adsorbed on oxygen-terminated single-crystal diamond (100), with a work function changing from 2.6 to 3.8 eV after successive steps of annealing at increasingly higher temperatures (the work function here is the energy difference between the Fermi- and vacuum-levels; since the Fermi level lies 1 eV above the top of the valence band, this range corresponds to $\varphi = 3.6 - 4.8$ eV). New techniques in growing nanocrystalline diamond (NCD) and ultrananocrystalline diamond (UNCD) [51], provide a high degree of flexibility in tuning the properties of the layer for a wide range of applications. In particular, hydrogen termination of nitrogen-incorporated 150 nm thick layer of UNCD ((N)UNCD:H) was shown to provide a stable work function of 3.1 eV, as well as protection against exposure to air [52]. A potential problem with UNCD, however, is a high degree of EFE [53-55], which may be prohibitive for large surface fields (tens of kV/cm); it remains to be demonstrated if a sufficiently strong EFE suppression is possible by tuning the UNCD parameters without, at the same time, paying a penalty of a higher work function. EFE is reported to be very low in single-crystal diamond [55], but it is unclear if this is a practical option for TPCs. The surface treatment of the above diamond layers provides a sufficiently high conductivity to avoid charging-up.

Mono- and bi-alkali photocathodes are perhaps the most sensitive IISEE materials, due to their exceedingly small work functions. At the same time, they are extremely sensitive to exposure to low levels of humidity and cannot be heated. Their use in rare-event TPCs, which must typically take data for several years, therefore seems highly challenging.

5. The case of $0\nu\beta\beta$ in ^{136}Xe

IOP Conf. Series: Journal of Physics: Conf. Series **1029** (2018) 012004 doi:10.1088/1742-6596/1029/1/012004

The concept of a high-pressure Xe (HPXe) TPC for $0\nu\beta\beta$ searches in ^{136}Xe, developed by the NEXT collaboration [56] and recently adopted by PandaX-III [57], has two unique features: excellent energy resolution (0.5-0.7% FWHM at $Q_{\beta\beta} = 2.46$ MeV), and track reconstruction capability. High energy resolution is critical for rejecting $2\nu\beta\beta$ events with energy close to $Q_{\beta\beta}$, while accurate imaging can provide effective discrimination against gamma-induced background events, based on the track topology. In the absence of a magnetic field, the main difference between a candidate ~20 cm long $0\nu\beta\beta$ track and a single-electron track is that the former has two 'blobs' of dense ionization at its opposite ends, while the latter generally has only one. The ability to extract topological information from ionization tracks was demonstrated in [58] using the NEXT-DEMO prototype, and the possibility of employing deep neural network algorithms to further enhance topology-based background rejection was demonstrated (on simulated tracks) in [59].

Current limits on $0\nu\beta\beta$ halflives indicate that the mass of future experiments should be on the ton-scale. The drift length L of a TPC containing 1 ton of Xe at 10 bar will be ~3 m (for equal diameter and length) or ~1.5 m in a bidirectional design. For a realistic field $E = 300$ V/cm the r.m.s. electron diffusion in pure Xe over 1.5 m drift is $\sigma_{el} \approx 16$ mm [2]. This will likely wash out the track blob structures, thus seriously affecting background rejection. Electron diffusion can be reduced by small admixtures of polyatomic gas, but this results in lower energy resolution. For example, simulations indicate that adding 0.5% CH_4 to pure Xe will reduce σ_{el} over 1.5 m from 16 to ~3 mm, but at the price of 1.6 larger FWHM [2].

Detection of the positive ions of a candidate $0\nu\beta\beta$ track *in addition* to its electrons could potentially enable accurate track imaging along with the excellent energy resolution of pure Xe. In this scheme (Figure 3) the TPC cathode plane is modified to detect positive ions, while the anode plane still includes an EL structure (e.g., two parallel meshes, or a THGEM-like structure). Both the cathode and anode planes are equipped with position-sensitive arrays of photon detectors. Primary scintillation is recorded on both planes, providing a clear t_0 signal. The electrons reach the anode plane within several hundred μs to produce an EL signal for precise measurement of the energy. A gross image of the track is recorded by the photon detectors at the anode plane. The ions reach the cathode several seconds later, where their position and time of arrival are recorded with high accuracy; the gross electron-based image recorded at the anode serves to narrow the data acquisition window for ion detection in terms of time and position.

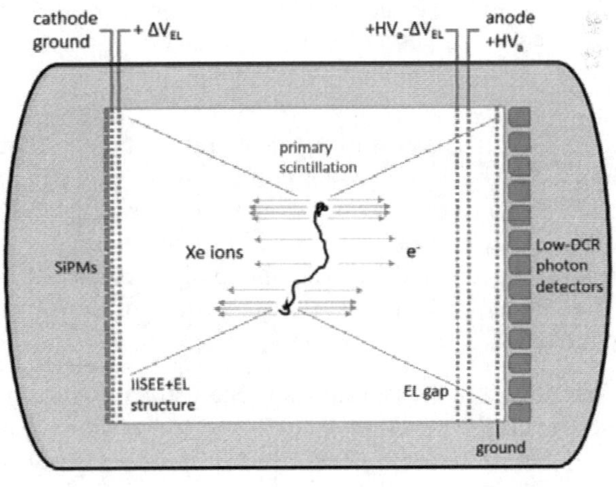

Figure 3: Proposed (one-directional) scheme for HPXe TPC for $0\nu\beta\beta$ searches in ^{136}Xe. Positive Xe ions are detected by ion-induced secondary electron emission (IISEE) followed by electroluminescence (EL) on the cathode plane, with accurate imaging by a dense array of SiPMs. Electrons produce an EL signal at the anode for energy measurement and gross imaging by the nearby photon detectors. Primary scintillation for t_0 determination is recorded on both planes (but more easily extracted from the low dark-count-rate detectors on the anode side). A bidirectional scheme where the design is mirrored with respect to a central cathode plane is also possible.

The cathode plane comprises two elements: an ion-induced EL structure and a dense array of pixelated detectors (e.g., silicon photomultipliers, SiPMs) closely behind. Two possible options are depicted in Figure 4: (A) a micro-strip gas counter (MSGC) pattern on tiled modules of fused silica plates, made up

IOP Conf. Series: Journal of Physics: Conf. Series **1029** (2018) 012004 doi:10.1088/1742-6596/1029/1/012004

of ion-collecting IISEE strips (e.g. MgO or NEA diamond on Mo) interlaced with electron-collecting metal strips; (B) two wire planes, where the one closer to the photon detectors (i.e., the cathode) includes Mo or W wires coated with MgO or NEA diamond, and the other serves to define the field. In (A), ions landing on the IISEE strips lead, with some probability, to the emission of secondary electrons; these are extracted into gas by the strong surface field, undergo modest avalanche multiplication, produce a localized pulse of EL and (in part) land on the neighboring metal strips; the same occurs in (B), where the emitted electrons cross the EL gap and are partly collected on the second wire plane. In both configurations the IISEE strips and wires can be ohmically flash-heated following a trigger based on t_0 and the anode EL signals; this will allow for precise timing of the flash-heating current pulse to bring the strips to a desired temperature when the ions arrive. To facilitate flash-heating and minimize the distance between the IISEE+EL structure and photon-detection plane, the cathode will be at ground potential, with the anode at +HV. This will require a ~10 cm gap between the anode EL structure and its photon-detector array (which should be further screened by a grounded mesh).

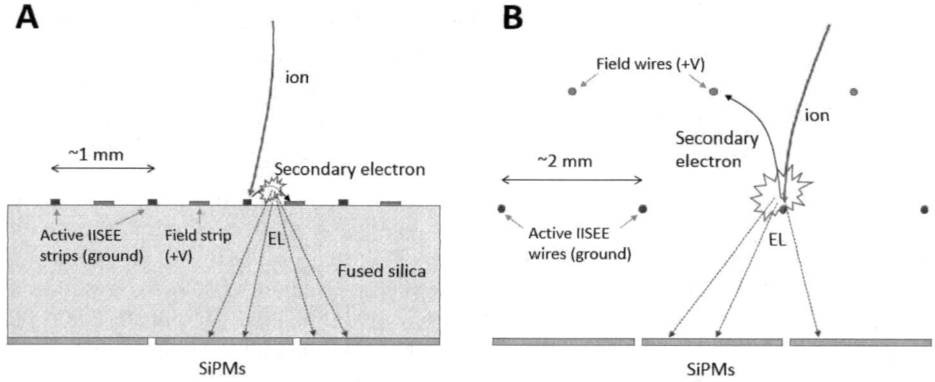

Figure 4: Possible ion detection structures on the cathode side (here drawn horizontally). (A) MSGC-like structure, with ion-collecting active IISEE strips interlaced with electron-collecting metal strips, deposited on a fused silica plate. Secondary electrons emitted by arriving ions from the active strips undergo modest avalanche multiplication and produce EL light - detected by a dense array of SiPMs – and are collected by the metal field strips. (B) The same principle, implemented with parallel planes of active IISEE wires and field wires. Dimensions are not to scale.

The photon detectors of the anode plane should probably not be SiPMs, to simplify the detection of the t_0 signal above the dark count noise; instead, one can consider a closely packed array of photomultiplier tubes (PMTs) or an array of large, highly-pixelated, gaseous photomultipliers (GPMs) behind fused silica windows [60,61]. Both options will provide a very low dark count rate (DCR). For GPMs fine granularity will likely be required for maintaining high energy resolution for the anode EL signal, by photon-counting on individual pixels; this is needed because of the exponential single-electron pulse height distribution of gas detectors.

We now turn to semi-quantitative estimates regarding different aspects of the proposed scheme. A candidate $0\nu\beta\beta$ event produces $\sim 1 \times 10^5$ electron-ion pairs which escape recombination [7]. The Xe^+ ions immediately (~0.1 ns) convert in three-body collisions to Xe_2^+ dimer ions [40], which may quickly convert to Xe_3^+ or larger clusters [41-44]. Xe neutral atom clustering around Xe ions is not sufficiently studied at 10 bar [7]; hopefully clustering results in one dominant ion species. If several cluster sizes coexist there will be a spread of ion drift velocities, smearing out the track along the drift direction; this, however, should not directly affect the reconstructed xy projection of the track. Extrapolating from measured values of the reduced mobility of Xe_3^+ at 0.07 bar [41], the expected drift velocity at 10 bar under 300 V/cm would be ~10-15 cm/s. Careful measurements of Xe cluster mobility at high pressure are obviously of vital importance here.

The ionization energy of Xe_2^+ is 11.18 eV [62]. Data for larger clusters is less clear, but it appears that the ionization energy is roughly constant regardless of the cluster size n for $n > 2$, with values around 10.5-11 eV [63, 64]. Since the gas is continuously purified, and since all common impurities (H_2, H_2O, O_2, N_2, CO_2, CO, CH_4) have ionization energies larger than 12 eV, the Xe ion clusters can be expected to reach the cathode without being neutralized by charge-exchange collisions along their drift.

From Einstein's relation, the r.m.s. diffusion of thermal ions (irrespective of their mass) is given by:

$$\sigma_{ions} = \sqrt{2Dt} = \sqrt{\frac{2k_BT}{q_e} \cdot \frac{L}{E}} = 2.25\sqrt{\frac{L}{E}} \tag{1}$$

where σ_{ions} is in mm, L in cm and E in V/cm. For $L = 150$ cm, $E = 300$ V/cm this gives $\sigma_{ions} =$ 1.6 mm, a 10-fold improvement compared to electrons (note that this is the *maximal* spread, for ions crossing the entire TPC). The spacing between the IISEE strips or wires should preferably be smaller than this to keep the track image as sharp as possible. For wire planes, the realistic distance between wires is probably limited to ~2 mm. For strips on fused silica the spacing can be smaller; however, going below ~1 mm between active strips (0.5 mm between active and electron-collecting field strips), may limit the EL yield unless substantial charge gain is allowed. Preliminary calculations which assume an overall photon detection efficiency of 10% for the cathode SiPM array (with 6 mm square pixels 3 mm from the strips) show that in order to reconstruct the ion arrival point with ~0.5 mm r.m.s. uncertainty, the EL pulse it induces should comprise ~2000 photons emitted into 4π (resulting in ~100 detected photons). Given the small distances between strips/wires this should be possible if the surface field is large enough to allow an amplification charge gain of ~5-10 (requiring ~200-400 photons per electron). Gain fluctuations seem unimportant as the amplification is not aimed at energy measurement. Careful optimization is required, but the numbers appear feasible with some flexibility in the geometry and applied voltages. The DCR of current generations of SiPMs at room temperature is $\sim 5 \times 10^4$ Hz/mm^2. If the EL signal lasts $\lesssim 100$ ns (where most of the light is emitted ~1 mm from the strip/wire), and 100 photons are detected over an active SiPM area of ~500 mm^2, the signal-to-noise ratio will be ~40. Care should be taken not to increase the SiPM temperature during flash-heating of the strips/wires. Alternatively, if DCR becomes an issue, one may consider modest cooling of the SiPM plane.

The main unknown in the proposed scheme is the IISEE yield for Xe ion clusters impinging on the active material. Since for molecular ions the yield is smaller than for atomic ions of similar ionization energy [32, 33], the expected yield for materials with an energy barrier for electron emission of ~4.3-4.7 eV (Mo, W, diamond with modest NEA) will likely be considerably smaller than 2%. This, however, may significantly improve for materials with $\varphi \sim 3-4$ eV (e.g., few atomic layers of MgO on Mo, or some UNCD variants). Since the strips/wires can be flash-heated upon trigger (and can remain at elevated temperature when the ions arrive), the presence of an adsorbed monolayer of either molecular impurities or physisorbed Xe atoms can be avoided. As noted above, care should be taken not to induce large convective flows during heating.

The probability of IISEE into the gas is the effective yield: $\gamma_{eff} = \gamma_i \cdot \varepsilon_{ext}(Xe; E/p, \bar{E}_k)$. Monte Carlo calculations on photoelectron emission from CsI into pure Xe, under irradiation at 185 nm (photon energy of 6.7 eV, corresponding to a maximum initial electron kinetic energy of ~0.6 eV [65]), show that $\varepsilon_{ext} \approx 0.2$ already at $E/p = 1\ \mathrm{Vcm^{-1}Torr^{-1}}$ (corresponding to 7.5 kV/cm at 10 bar) [66]. (The simulations were validated experimentally on Xe/CH_4 mixtures down to 1% CH_4.) Preliminary Garfield++ [67] Monte Carlo calculations of the extraction efficiency into Xe at 10 bar are consistent with

[66] and indicate that it increases significantly at higher fields; these calculations further show that for a given field ε_{ext} decreases with the initial kinetic energy of the emitted electron.

In order to accurately reconstruct the topology of a candidate $0\nu\beta\beta$ event it appears that several hundred detected ions are sufficient. For example, if 200 ions are detected, the typical spacing between adjacent ions for a 20 cm long track will be ~1 mm, with higher densities at the opposite ends of the track. For 1×10^5 ions arriving at the cathode, this will require $\gamma_{eff} = 0.2\%$ (e.g. by having, $\gamma_i = 1\%$ and $\varepsilon_{ext} = 0.2$). Higher values for γ_{eff} will obviously improve the quality of the image (which will inevitably contain some noise). Note that in addition to identifying the blob structures, further information on the track curvature (which should be readily accessible with $\gtrsim 200$ detected ions) may assist in background discrimination [68].

The main source of noise will be random electron field emissions from the IISEE strips/wires. Assuming 1 detected ion per mm^3 of track, and a drift velocity of ~10 cm/s, an overall EFE rate of 10 Hz/mm^2 (over the cathode surface) will result in a signal-to-noise ratio of 10. If the cathode plane is made up of 20 μm-wide strips with a spacing of 1 mm, this corresponds to an EFE rate of 500 Hz per mm^2 of active IISEE material (into Xe at 10 bar). Assuming $\varepsilon_{ext} = 0.2$ this translates in vacuum to EFE current density of $4 \times 10^{-14} A/cm^2$ of active IISEE material. This appears to be a challenging requirement of UNCD [53-55], but may possibly be overcome by tuning the layer's parameters. As noted above, there are no known data on EFE from ultra-thin MgO on Mo.

Photon feedback from the IISEE strips/wires may potentially become an issue if the quantum efficiency of the active material at 172 nm is large. However, the risk of uncontrolled feedback can be mitigated by optimizing the geometry and voltages of the IISEE-EL structure – e.g., by reducing the ratio between the strip width and spacing, by reducing the surface field on the active strips and by allowing part of the avalanche to develop next to the collecting strips (with similar consideration for wire planes).

6. Concluding remarks

Potential emission of secondary electrons by positive ions may open new experimental avenues in low-rate gaseous TPCs. Since the basic mechanism can work for thermal ions, it may be applied in TPCs operating at atmospheric or high pressure. The intrinsic spatial resolution is governed by thermal diffusion and can be on the sub-mm scale, depending on the TPC size and field strength. In principle, the proposed method can allow the detection of individual ions, one-by-one. The key requirements from the active materials are a low energy barrier for electron emission, high degree of chemical inertness and/or suitability for *in situ* heating, and a low rate of electron field emission.

Considering the case of $0\nu\beta\beta$ searches in ^{136}Xe using a HPXe TPC, positive ion detection may allow ~1-2 mm accuracy in track reconstruction without compromising the excellent energy resolution of pure Xe. It appears that the main challenge will be identifying an electron emitting material that offers an IISEE yield of a few percent for Xe ion clusters, accompanied by very low electron field emission. Luckily, there appear to be viable materials that can serve as a starting point for this search. A potential complication is the possibility that several sizes of Xe ion clusters coexist with similar concentrations at high pressure. This may interfere with track reconstruction along the drift direction, but should not directly affect the accuracy in *xy*. Other R&D issues include the (likely) necessity to flash-heat the IISEE strips/wires following a trigger from t_0 and the anode EL signal, and the multivariate design of the IISEE+EL structure and readout. Finally, realistic estimates should be made regarding the impact of the proposed scheme on the overall sensitivity of the experiment for $0\nu\beta\beta$ detection.

Conceivably, as noted by D. Nygren in [7] with respect to $0\nu\beta\beta$, and by N. Spooner with respect to directional dark matter searches [69], one may consider incorporating positive ion detection into negative ion TPCs. Considering the efficiency of positive ion detection, the potential merit of this will be the addition of an 'echo' signal originating from electron release from the cathode when the ions arrive, enabling to position the event in three dimensions.

Experimental measurements of IISEE yields and field emission rates of candidate materials, with particular emphasis on Xe, are now underway. As a directly relevant byproduct, these experiments are expected to provide valuable information on ion cluster mobilities at different pressures.

The challenges associated with harnessing ion-induced potential emission to gaseous TPCs are varied and numerous, and addressing them will likely unravel new ones. Yet, in view of the rapid advances in material and surface science, one can hope that this idea may eventually become a reality.

Acknowledgements

I would like to thank A. Breskin for his ongoing support and encouragement, as well as for introducing me to past works of relevance to this proposal; D. Nygren for the stimulating discussions that eventually led to this idea, and for many insightful comments along the way; S. Shchemelinin for eye-opening discussions on potential emission; A. Para for the enthusiastic Ping-Pong of comments and ideas and for initiating the connection with ANL, and A. Sumant and A. Hoffman for the discussions on diamond layers and for their willingness to join the experimental effort. This work is supported by the Weizmann Institute Staff Scientist Internal Grant Program.

References

[1] Fancher D *et al.* 1979 Performance of a time-projection chamber *Nucl. Instrum. Meth.* A **161** 383-90

[2] Azevedo C D R, Fernandes L M P, Freitas E D C, Gonzalez-Diaz D, Monrabal F, Monteiro C M B, Dos Santos J M F, Veloso J F C A and Gomez-Cadenas J J 2016 An homeopathic cure to pure Xenon large diffusion, 2016 *JINST* **11** C02007

[3] Martoff C J, Snowden-Ifft D P, Ohnuki T, Spooner N and Lehner M 2000 Suppressing drift chamber diffusion without magnetic field *Nucl. Instrum. Meth.* A **440** 355-9

[4] Battat J B R *et al.* (DRIFT collaboration) 2015 First background-free limit from a directional dark matter experiment: results from a fully fiducialised DRIFT detector *Physics of the Dark Universe* **9–10** 1–7

[5] Prieskorn Z R, Hill J E, Kaaret P E and Black J K 2014 Photoelectron track length distributions measured in a negative ion time projection chamber *IEEE Trans. Nucl. Sci.* **61** 894-900

[6] Sauli F 2004 Progress with the gas electron multiplier *Nucl. Instrum. Meth.* A **522** 93-8

[7] Nygren D R 2007 A negative-ion TPC with ultra-high energy resolution for 0-ν double beta decay search in ^{136}Xe *J. Phys.: Conf. Ser.* **65** 012003

[8] Chmelewski D, Parmentier N, Le Grand J 1974 Dispositif experimental en vue d'etudes dosimetriques au niveau du nanometer *Proc. 4th Symp. on Microdosimetry, EUR 5122 d-e-f, CEC (Luxembourg)* p 869

[9] Pszona S 1975 A track ion counter *Proc. 5th Symp. on Microdosimetry, EUR 3452 d-e-f, CEC (Luxembourg)* p 1107

[10] Shchemelinin S, Breskin A, Chechik R, Pansky A, Colauttib P, Conteh V, De Nardob L and Torniellib G 1996 Ionization measurements in small gas samples by single ion counting *Nucl. Instrum. Meth.* A **368** 859-61

[11] Garty G, Shchemelinin S, Breskin A, Chechik R, Assaf G, Orion I, Bashkirov V, Schulte R and Grosswendt B 2002 The performance of a novel ion-counting nanodosimeter *Nucl. Instrum. Meth.* A **492** 212–35

[12] Pszona S, Bantsar A and Nikjoo H 2006 Ionization cluster size distribution for alpha particles: experiment, modelling *Rad. Prot. Dos.* **122** 28–31

[13] Garty G, Schulte R, Shchemelinin S, Leloup C, Assaf G, Breskin A, Chechik R, Bashkirov V, Milligan J and Grosswendt B 2010 A nanodosimetric model of radiation-induced clustered DNA damage yields *Phys. Med. Biol.* **55** 761–81

[14] Breskin A, Alon R, Cortesi M, Chechik R, Miyamoto J, Dangendorf V, Maia J and Dos Santos J M F 2009 A concise review on THGEM detectors *Nucl. Instr. Meth.* A **598** 107-11

[15] Bashkirov V, Hurley R F and Schulte R W 2009 A novel detector for 2D ion detection in low-pressure gas and its applications *IEEE Nucl. Sci. Symp. Conf. Rec. (NSS/MIC)* 694.

[16] Vasi F, Casiraghi M, Bashkirov V, Giesen U and Schulte R W 2016 Development of a single ion detector for radiation track structure studies 2016 *JINST* **11** C09021

[17] Shchemelinin S and Breskin A 2017 Observation of electron excitation into silicon conduction band by slow-ion surface neutralization 2017 *JINST* **12** P03020

[18] Dangendorf V, private communication

[19] Hagstrum H D 1954 Theory of Auger ejection of electrons from metals by ions *Phys. Rev.* **96** 336-65

[20] Baragiola R A 1982 Principles and applications of ion-induced Auger electron emission from solids *Radiat. Eff.* **61** 47-72

[21] Varga P and Winter H 1991 Slow particle-induced electron emission from solid surfaces *Particle Induced Electron Emission II (Springer Tracts in Modern Physics* vol 123*)* ed D Hasselkamp *et al* (Berlin: Springer-Verlag) pp 149-214

[22] Zeijlmans van Emmichoven P A and Matulevich Y T 2007 Slow ion-induced electron emission from thin insulating films *Slow Heavy-Particle Induced Electron Emission from Solid Surfaces (Springer Tracts in Modern Physics* vol 225*)* ed H P Winter and J Burgdörfer (Berlin: Springer) pp 213-39

[23] Baragiola R A and Riccardi P 2008 Electron emission from surfaces induced by slow ions and atoms *Reactive Sputter Deposition (Springer Series in Materials Science* vol 109*)* ed D Depla and S Mahieu (Berlin: Springer-Verlag) pp 43-60

[24] Carmina Monreal R 2014 Auger neutralization and ionization processes for charge exchange between slow noble gas atoms and solid surfaces *Prog. Surf. Sci.* **89** 80–125

[25] Hagstrum H D 1954 Auger ejection of electrons from tungsten by noble gas ions *Phys. Rev.* **96** 325-35

[26] Hagstrum H D 1956 Auger ejection of electrons from molybdenum by noble gas ions *Phys. Rev.* **104** 672-83

[27] Hagstrum H D 1956 Effect of monolayer adsorption on the ejection of electrons from metals by ions *Phys. Rev.* **104** 1516-27

[28] Hagstrum H D 1960 Auger electron ejection from germanium and silicon by noble gas ions *Phys. Rev.* **119** 940-56

[29] Hagstrum H D 1961 Theory of Auger neutralization of ions at the surface of a diamond-type semiconductor *Phys. Rev.* **122** 83-113

[30] Hagstrum H D 1966 Ion-Neutralization Spectroscopy of solids and solid surfaces *Phys. Rev.* **150** 495-515

[31] A V Phelps and Petrović Z Lj 1999 Cold-cathode discharges and breakdown in argon: surface and gas phase production of secondary electrons *Plasma Sources Sci. Technol.* **8** R21–R44

[32] Vance D W 1968 Auger electron emission from clean Mo bombarded by positive ions. II. Effect of angle of incidence *Phys. Rev.* **169** 252-62

[33] Vance D W 1968 Auger electron emission from clean Mo bombarded by positive ions. III. Effect of electronically excited ions *Phys. Rev.* **169** 263-72

[34] Motoyama Y and Sato F 2006 Calculation of Secondary Electron Emission Yield γ From MgO Surface *IEEE Trans. on Plasma Sci.* **34** 336-42

[35] Matulevich Y T, Vink T J and Zeijlmans van Emmichoven P A 2002 Low-energy ion-induced electron emission from a MgO(100) thin film: the role of the MgO-substrate interface *Phys. Rev. Lett.* **89** 167601

[36] Matulevich Y T and Zeijlmans van Emmichoven P A 2004 Electron emission from low-energy Xe+ ions interacting with a MgO thin film deposited on a Mo substrate *Phys. Rev.* B **69** 245414

[37] van Elsbergen V, Bachmann P K and Juestel T 2000 Ion-induced secondary electron emission: a comparative study *Conf. Soc. of Information Display, SID00 Digest* 220

[38] Bachmann P K, van Elsbergen V, Wiechert D U, Zhong G and Robertson J 2001 CVD diamond: a novel high γ-coating for plasma display panels? *Diam. Relat. Mater.* **10** 809-17

[39] Lyashenko A, Breskin A, Chechik R and Dias T H V T 2009 Ion-induced secondary electron emission from K–Cs–Sb, Na–K–Sb, and Cs–Sb photocathodes and its relevance to the operation of gaseous avalanche photomultipliers *J. Appl. Phys.* **106** 044902

[40] Neves P N B, Conde C A N and Távora L M N 2010 The $X^{+} + 2X \to X_2^{+} + X$ reaction rate constant for Ar, Kr and Xe, at 300 K *Nucl. Instrum. Meth.* **A** 619 75–7

[41] Helm H 1976 Formation of Xe_3^{+} ions in xenon at temperatures between 210 and 293 K *Phys. Rev.* A **14** 680-4

[42] Amarouche M, Durand G and Malrieu J P 1988 Structure and stability of Xe_n^{+} clusters *J. Chem. Phys.* **88** 1010-8

[43] Hiraoka K and Mori T 1990 Stability of rare gas cluster ions *J. Chem. Phys.* **92** 4408-4416

[44] Itoh K, Holroyd R A and Nishikawa M 2001 Ion Mobilities in Supercritical Ethane, Xenon, and Carbon Dioxide *J. Phys. Chem.* A **105** 703-9

[45] Kalkan Y, Arslandok M, Cortez A F V, Kaya Y, Tapana I and Veenhof R 2015 Cluster ions in gas-based detectors 2015 *JINST* **10** P07004

[46] Escada J, Dias T H V T, Rachinhas P J B M, Santos F P, Lopes J A M, Coelho L C C, Conde C A N and Stauffer A D 2010 A Monte Carlo study of photoelectron extraction efficiency from CsI photocathodes into Xe–CH_4 and Ne–CH_4 mixtures *J. Phys. D: Appl. Phys.* **43** 065502

[47] Stuckenholz S, Büchner C, Heyde M and Freund H J 2015 MgO on Mo(001): Local work function measurements above pristine terrace and line defect sites *J. Phys. Chem.* C **119** 12283–90

[48] Giordano L, Cinquini F and Pacchioni G 2005 Tuning the surface metal work function by deposition of ultrathin oxide films: Density functional calculations *Phys. Rev.* **B** 73 045414

[49] Michaelson Sh, Akhvlediani R, Ternyak O, Hoffman A, Breskin A and Chechik R 2005 Absolute quantum photo-yield of nanometer thick diamond films at their initial stages of formation *Diam. Relat. Mater.* 14 546– 51

[50] O'Donnell K M, Edmonds M T, Tadich A, Thomsen L, Stacey A, Schenk A, Pakes C I and Ley L 2015 Extremely high negative electron affinity of diamond via magnesium adsorption *Phys. Rev.* **B** 92, 035303

[51] Butler J E and Sumant A V 2008 The CVD of Nanodiamond Materials *Chem. Vap. Deposition* **14** 145–60

[52] Pérez Quintero K J, Antipov S, Sumant A V, Jing Ch and Baryshev S V 2014 High quantum efficiency ultrananocrystalline diamond photocathode for photoinjector applications *Appl. Phys. Lett.* **105** 123103

[53] A. R. Krauss *et al* 2001 Electron field emission for ultrananocrystalline diamond films *J. Appl. Phys.* **89** 2958-67

[54] Joseph P T, Tai N H, Chen C H, Niu H, Cheng H F, Pong W F and Lin I N 2009 On the mechanism of enhancement on electron field emission properties for ultrananocrystalline diamond films due to ion implantation *J. Phys. D: Appl. Phys.* **42** 105403

[55] Harniman R L, Fox O J L, Jansse W, Drijkoningen S, Haenen K and May P W 2015 Direct observation of electron emission from grain boundaries in CVD diamond by PeakForce-controlled tunneling atomic force microscopy *Carbon* **94** 386–95

IOP Conf. Series: Journal of Physics: Conf. Series **1029** (2018) 012004 doi:10.1088/1742-6596/1029/1/012004

[56] The NEXT collaboration (Gómez Cadenas J J *et al.*) 2014 Present Status and Future Perspectives of the NEXT Experiment *Adv. High Energy Phys.* **2014** 907067

[57] Chen X *et al* 2016 PandaX-III: Searching for neutrinoless double beta decay with high pressure ^{136}Xe gas time projection chambers, arXiv:1610.08883

[58] The NEXT Collaboration (P. Ferrario *et al.*) 2016 First proof of topological signature in the high pressure xenon gas TPC with electroluminescence amplification for the NEXT experiment *JHEP* **1601** 104

[59] The NEXT Collaboration (J. Renner *et al.*) 2017 Background rejection in NEXT using deep neural networks 2017 *JINST* **12** T01004

[60] Chechik R and Breskin A 2008 Advances in gaseous photomultipliers *Nucl. Instrum. Meth.* A **595** 116-27

[61] Arazi L, Coimbra A E C, Erdal E, Israelashvili I, Rappaport M L, Shchemelinin S, Vartsky D, dos Santos J M F and Breskin A 2015 First results of a large-area cryogenic gaseous photomultiplier coupled to a dual-phase liquid xenon TPC 2015 *JINST* **10** P10020

[62] Lu Y, Morioka Y, Matsui T, Tanaka T, Toshii H, Hall R I, Hayaishi T and Ito K 1995 Ground and excited states of Xe_2^+ observed by high resolution threshold photoelectron spectroscopy of Xe_2 *J. Chem. Phys.* **102** 1553-1560

[63] Echt O 2004 Ionization energies of van der Waals clusters, presentation in EPIC Meeting, Obergurgl, Austria June 18 – 23, 2004

[64] Echt O, Fiegele T, Rümmele M, Probst M, Matt-Leubner S, Urban J, Mach P, Leszczynski J, Scheier P and Märk T D 2005 Ionization energies of argon clusters: A combined experimental and theoretical study *J. Chem. Phys.* **123** 084313

[65] Breskin A 1996 CsI UV photocathodes: history and mystery *Nucl. Instrum. Meth.* A **371** 116-36

[66] Coelho L C C, Lopes J A M, Escada J, Dias T H V T, dos Santos J M F 2009 Photoelectron transmission efficiency in Ar–CH_4 and Xe–CH_4 mixtures: Experimental results *Nucl. Instrum. Meth.* A **607** 587-90

[67] H. Schindler, Garfield++. http://garfieldpp.web.cern.ch/garfieldpp/, Accessed: 21-03-2017

[68] A. Para, private communication

[69] N. Spooner, private communication

IOP Conf. Series: Journal of Physics: Conf. Series **1029** (2018) 012005 doi:10.1088/1742-6596/1029/1/012005

Dark Matter Directional Detection with MIMAC

Daniel Santos
LPSC, Université Grenoble-Alpes, CNRS/IN2P3, 53, Av. des Martyrs, 38026 Grenoble, France
E-mail: daniel.santos@lpsc.in2p3.fr

Abstract. The MIMAC (MIcro-tpc MAtrix of Chambers) experiment is a micro-TPC matrix for directional dark matter search. Directional detection is a strategy based on the measurement of the WIMP flux anisotropy due to the solar system motion with respect to the dark matter halo. The main purpose of MIMAC project is the measurement of nuclear recoils energy and their 3D track direction from the WIMP elastic scattering on target nuclei. The first ionization energy and 3D track observations of nuclear recoils produced by the radon progeny recently reported being shown. This measurement shows the capability of the MIMAC detector and opens the possibility to explore the low energy recoil directionality signature. The measurement of the drift velocity from an implementation of the cathode signal is also mentioned.

1. Introduction

A wide variety of astrophysical observations at different scales converge to the hypothesis of the existence of an important dark matter component in our Universe. According to cosmological precision measurement, dark matter represents roughly 85% of the matter in the Universe [1]. At the galaxy scale, the dark matter particles would form cold and static halos around the disks of galaxies. This dark halo would be composed by particles called WIMP (Weakly Interacting Massive Particle), a stable particle interacting only by weak and gravitational interactions with a mass ranging from 1 GeV to 1 TeV scale. The motion of the solar system with respect to the dark halo should generate a detectable WIMP flux through the WIMP-nucleus interaction on an underground detector. The dark matter directional detection opens a new field in cosmology bringing the possibility to build a map of nuclear recoils that would be able to explore the galactic dark matter halo giving access to a particle characterization of such matter and the shape of the halo [10]. The directional signature would be the only way to go further the ultimate neutrino background [11]. MIMAC is a dark matter directional detection project based on a matrix of micro-TPC chambers at low pressure having a high spatial resolution in 3D [2, 4]. As the other directional detection experiments [3], the aim of the MIMAC project is the measurement of the WIMP-induced nuclear recoil energy and angular distributions in order to constrain the WIMP dark matter properties. In this context, the MIMAC collaboration developed an original detector and readout strategy allowing the measurement of nuclear recoil tracks in 3D. Since June 2012, a bi-chamber prototype was installed at the Modane underground laboratory (Laboratoire Souterrain de Modane[1]) for preliminary tests. This first data analysis showed the most important source of background for dark matter search came from the radon progeny recoils. The first 3D tracks measurement of the daughter nuclear recoils from radon progeny α-decay were detected and recently published [13].

[1] LSM

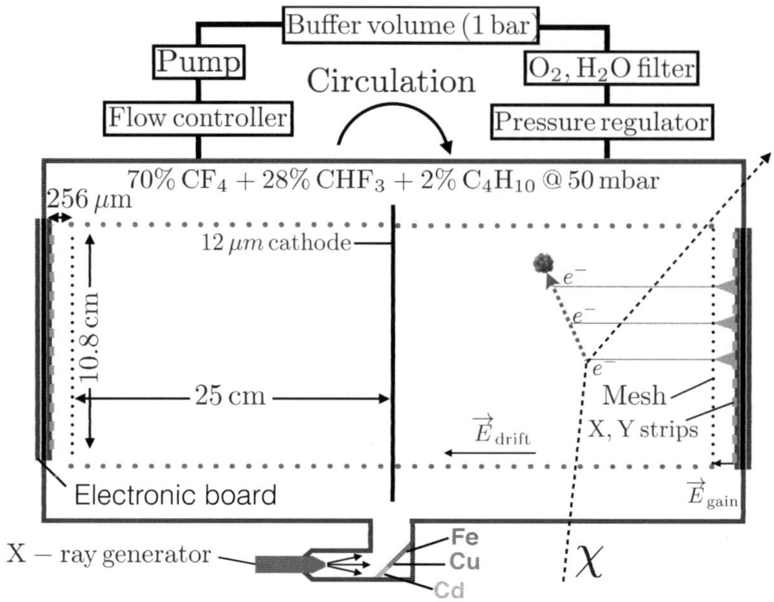

Figure 1. Scheme of the bi-chamber prototype configuration and the ionization electron collection from a nuclear recoil produced by a WIMP elastic scattering.

2. The MIMAC experiment at Modane Underground Laboratory (LSM)

The bi-chamber prototype is filled with a low pressure (50 mbar) $CF_4 + 28\%CHF_3 + 2\%C_4H_{10}$ gas mixture. As schematically shown in figure 1, this prototype is composed of two mirroring chambers sharing a thin 6 μm aluminized mylar cathode at 25 cm from the anode. The anode consists of a pixelated micromegas [5, 6] of 10.8 cm side linking the 200 μm pixels by strips with a 424 μm pitch coupled to a fast self-triggered electronic system specially developed for MIMAC [7, 8].

3. Radon Progeny spectra and their 3D tracks

The intrinsic pollution of ^{238}U and ^{232}Th of the detector materials is the source of ^{222}Rn and ^{220}Rn emanations. After successive alpha decays, radon isotopes can migrate from the materials bulk to their surfaces. The events associated with α and β decays of the radon chain radio-nuclei constitute a background for WIMP dark matter search and they should be discriminated in dark matter detectors. Radon Progeny Recoils (RPR) denote nuclear recoils produced by α-decays of radioactive elements of ^{222}Rn and ^{220}Rn decay chains, it includes α-particles and daughter nuclei recoils. In the MIMAC detector, in the dark matter search configuration, it is not possible to measure α-particle energies or their track lengths because at 50 mbar their tracks are not fully contained in the active volume of one chamber. Instead, we can observe the daughter nuclear recoils produced by the α-particle emission measuring their 3D tracks with their total ionization energy showing the ability of the detector to get a clear signature of low energy nuclear recoil tracks.

The ratios of the two peak amplitudes are 2.6 ± 0.3 for the chamber 1 and 2.4 ± 0.3 for the chamber 2. In addition, RPR events occur at different positions inside the detector which impact their tracks and energies and in consequence their discrimination. In figure 3 a RPR event of 41.1 keV is shown at the cathode level and another of 36.4 keV on a grid level. The difference between them is the 25 cm diffusion in their charge collection.

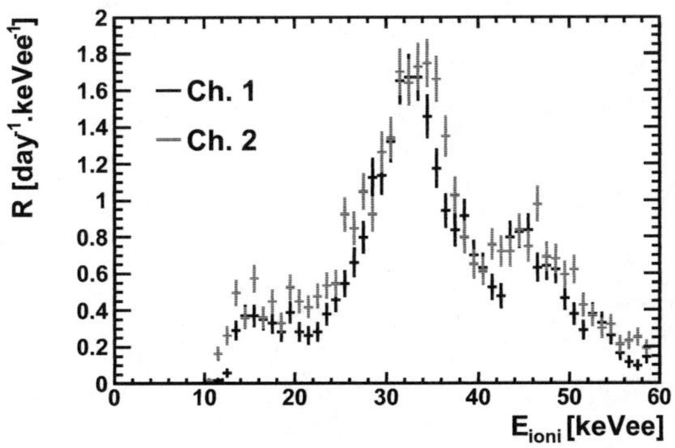

Figure 2. Energy spectra measured by the chamber 1 (black line) and the chamber 2 (red line). It should be pointed out that each event represented on these spectra has its own 3D track associated. These distributions were obtained applying the low energy electron/recoil discrimination.

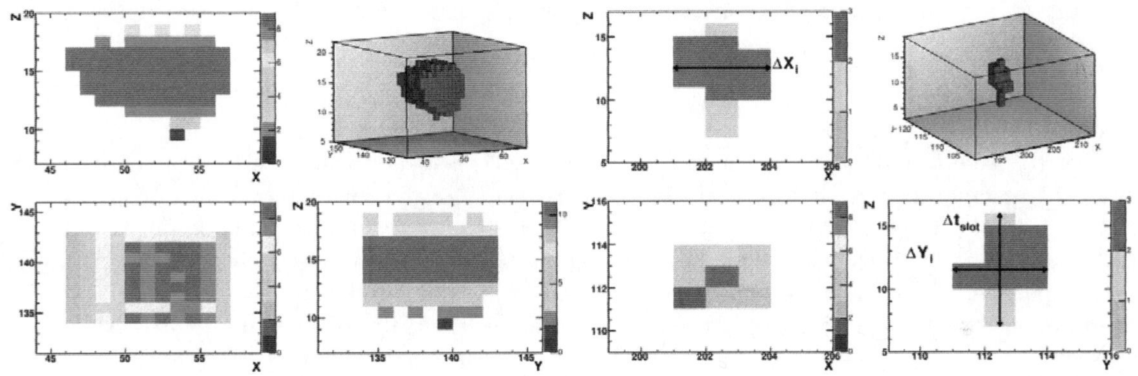

Figure 3. Projections of a 41.1 (left panels) and a 36.4 keVee (right panels) measured nuclear recoil tracks in the (X,Z), (Y,Z) and (X,Y) planes and 3D reconstructions. The Z axis is in units of time-slice (20 ns) and the X and Y axis in strip number. The color scale corresponds to the number of strips fired on the time-slice. The horizontal arrows represent one time-slice width X/Y along the X/Y axis. The vertical arrow represents the slot duration Δt_{slot}.

4. Electron-nuclear recoil discrimination

In order to reject the electron recoil background from dark matter searches data, we have developed an original electron/recoil discrimination method described in [15]. We placed a MIMAC mono-chamber detector on a monochromatic neutron field allowing us to acquire two specific data sets: electron and nuclear recoil and electron recoil only. We applied a Boosted Decision Tree (BDT) algorithm on these two data sets. It gives a 10^5 electron recoil rejection power. The application of the BDT analysis on the Monte Carlo shows a $86.49 \pm 0.17\%$ nuclear recoil efficiency considering the full energy range and $94.67 \pm 0.19\%$ considering a 5 keV energy threshold with a 10^5 electron recoil rejection power.

This method was applied on the 2013 data run (103 days) and figure 2 presents the resulting nuclear recoil energy spectra. After the application of this analysis, we can consider the electron recoil event contamination on our data as negligible. These energy spectra show two peaks at 33 and 46 keVee, in both chambers (black line chamber 1 and red line chamber 2) [13].

IOP Conf. Series: Journal of Physics: Conf. Series **1029** (2018) 012005 doi:10.1088/1742-6596/1029/1/012005

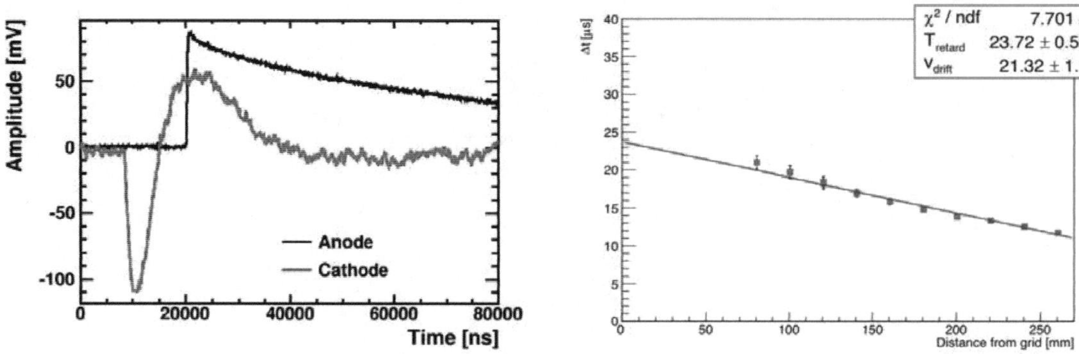

Figure 4. Left: amplitudes of the signals from the cathode preamplifier and from the grid preamplifier as shown on the oscilloscope. Note that here, the cathode signal is not delayed, and is detected before the grid signal. Right: Measure of the time differences (ΔTAC) between the grid signal and the delayed cathode signal as a function of the distance of the α source from the anode (green points) ; only the statistical errors are displayed. A linear fit of these points is superimposed in red and provides the values of the drift velocity and the imposed delay on the cathode signal.

5. The cathode signal

In order to place the 3D track in the active volume with respect to the position of the cathode, we have implemented an electronic device to take the induced current on the cathode signal produced by the movement of the primary ionization electrons during their transport to the grid by the drift electric field. In figure 4 we show the cathode signal generated by an alpha particle parallel to the cathode placed at 14 cm from it. The validation of this signal as a new observable has been done performing the measurement of the drift velocity in our MIMAC mixed gas chamber, see figure 4 and [14] for more details. The final value of the drift velocity obtained was:

$$v_{\text{drift}} = (21.2 \pm 1.2(\text{stat}) \pm 0.04(\text{sys}))\mu\text{m/ns}.$$

It differs slightly from the value obtained using the MAGBOLTZ software [16]:

$$v_{\text{drift}} = (24.1 \pm 0.02)\mu\text{m/ns}.$$

The difference can be explained by the experimental conditions, in general, not the same as those introduced as input in the simulation: the pressure of the mixed gas varying with the temperature of the room ($\sim$1 mbar), the gas containing some impurities not at all evaluated, and the electric field in the drift volume can contain some amount of inhomogeneity. The same difference has been reported by an alternative experimental method in 2013 [9]. The measurement of the drift velocity performed with the method described here corresponds to a setup with real experimental conditions, which are not fully taken into account when performing the MAGBOLTZ simulation.

6. Conclusions

In this proceeding, we have summarized some of the main results of MIMAC collaboration in the last years. We have shown, for the first time, the observation of low energy nuclear recoil 3D tracks from daughter nuclei of the ^{222}Rn decay chain. This measurement shows the capability of the MIMAC detector and opens the possibility to explore the low energy recoil directionality signature. The radon gas emanation monitoring and its reduction is a major topic for rare event experiments such as dark matter or double-β decay search experiments. The new degrees of freedom, offered by the observation of 3D low energy nuclear recoil tracks describing these events shed a new light on them improving their localization and discrimination. The implementation of the cathode signal in a TPC is an interesting result opening the possibility to place the 3D tracks in the active volume.

Acknowledgements
We thank the Laboratoire Souterrain de Modane (CNRS/CEA) and its team for their support all along our operation. MIMAC collaboration acknowledges the Labex Enigmass funding.

References

[1] P. A. R. Ade, N. Aghanim, M. Arnaud, M. Ashdown, J. Aumont, C. Baccigalupi et al., *Planck 2015 results*, *Astronomy & Astrophysics* **594** (oct, 2016) A13, [1502.01589].
[2] D. Santos, O. Guillaudin, T. Lamy et al., *MIMAC: A Micro-TPC Matrix of Chambers for direct detection of Wimps*, *J. Phys. Conf. Ser.* **65** (Apr., 2007) 012012, [0703310v1].
[3] S. Ahlen, N. Afshordi, J. Battat et al., *The case for a directional dark matter detector and the status of current experimental efforts*, *Int.J.Mod.Phys.* **A25** (2010) 1–51, [0911.0323].
[4] D. Santos, J. Billard, G. Bosson et al., *MIMAC : A micro-tpc matrix for dark matter directional detection*, *J. Phys. Conf. Ser.* **460** (Oct., 2013) 012007, [1304.2255].
[5] F. J. Iguaz, D. Attié, D. Calvet et al., *Micromegas detector developments for Dark Matter directional detection with MIMAC*, *J. Instrum.* **6** (July, 2011) P07002–P07002, [1105.2056].
[6] I. Giomataris, R. De Oliveira, S. Andriamonje et al., *Micromegas in a bulk*, *Nucl.Instrum.Meth.* **A560** (May, 2006) 405–408.
[7] J. Richer, G. Bosson, O. Bourrion et al., *Development of a front end ASIC for Dark Matter directional detection with MIMAC*, *Nucl.Instrum.Meth.* **A620** (2010) 470–476, [0912.0186].
[8] O. Bourrion, G. Bosson, C. Grignon et al., *Data acquisition electronics and reconstruction software for directional detection of Dark Matter with MIMAC*, *Nucl.Instrum.Meth.* **A662** (2010) 207–214, [1006.1335].
[9] J. Billard, F. Mayet, G. Bosson et al., *In situ measurement of the electron drift velocity for upcoming directional Dark Matter detectors*, *J. Instrum.* **9** (Jan., 2014) P01013–P01013, [1305.2360].
[10] F. Mayet et al. *A review of the discovery reach of directional Dark Matter detection Physics Reports* **627** (2016) 1, [1602.03781].
[11] J. Billard, L. Strigari and E. Figueroa-Feliciano, *Implication of neutrino backgrounds on the reach of next generation dark matter direct detection experiments*, *Phys.Rev.* **D89** (2014) 023524, [1307.5458].
[12] J. Billard, F. Mayet, J. Macias-Perez and D. Santos, *Directional detection as a strategy to discover galactic Dark Matter*, *Phys.Lett.* **B691** (2010) 156–162, [0911.4086].
[13] Q. Riffard et al.*First detection of radon progeny recoil tracks by MIMAC, JINST, 2017* **12, P06021**
[14] C. Couturier, Q. Riffard, N. Sauzet et al. *Cathode signal in a TPC directional detector: implementation and validation measuring the drift velocity, to be published in JINST 2017*
[15] Q. Riffard et al., *MIMAC low energy electron-recoil discrimination measured with fast neutrons, JINST, 2016* **11, P08011** 1602.01738.
[16] S. Biagi, *Monte Carlo simulation of electron drift and diffusion in counting gases under the influence of electric and magnetic fields*, *Nucl.Instrum.Meth.* **A421** (Jan., 1999) 234–240.

IOP Conf. Series: Journal of Physics: Conf. Series **1029** (2018) 012006 doi:10.1088/1742-6596/1029/1/012006

Spherical Proportional Counter: A review of recent developments

Q. Arnaud[1], A. Brossard[1], A. Dastgheibi-Fard[3], G. Gerbier[1], I. Giomataris[2], I. Katsioulas[2], P. Knights[2,4], K. Nikolopoulos[4] and I. Savvidis[5], on behalf of the NEWS-G collaboration (news-g.org)

[1] Department of Physics, Engineering Physics & Astronomy, Queens University, Kingston, Ontario K7L 3N6, Canada
[2] IRFU, CEA, Universite Paris-Saclay, F-91191 Gif-sur-Yvette, France
[3] LSM, CNRS/IN2P3, Universite Grenoble-Alpes, Modane, France
[4] School of Physics and Astronomy, University of Birmingham, B15 2TT, United Kingdom
[5] Department of Physics, Aristotle University of Thessaloniki, GR-52124 Thessaloniki, Greece

E-mail: `ioannis.katsioulas@cea.fr`

Abstract. A review of the key developments in the Spherical Proportional Counter is presented. The detector technology and operation principles are described along with results, such as the low-energy calibration, and more recent advances, including the use of resistive materials and a multi-ball readout system. The Spherical Proportional Counter has been utilised by the NEWS-G experiment, performing a direct search for light DM candidates, and a review of the recent results is provided. Prospects for future applications of the technology are also discussed.

1. Introduction

The Spherical Proportional Counter (SPC), was developed in 2006 at CEA Saclay by a group led by I. Giomataris [1, 2]. Originally it aimed at measuring "room" size oscillations of neutrinos from a ^{3}H source at its centre [3]. Apart from providing 4π coverage around the source, a spherical detector has the advantage that it contains the maximum volume for a given surface. Since the majority background in rare event detection experiments originates from the detector surface, such a detector would also maximize the signal to noise ratio.

The first prototypes were realised using decommissioned RF cavities from LEP with a diameter of approximately 1.3 m. The new detector, which has a simple and robust design, demonstrated impressive attributes such as very low energy threshold down to single electron detection, very low electronic noise even for large volumes, read-out through a single channel, and possibility to perform fiducial selections

The SPC can detect a wide range of radiations including nuclear recoils arising from elastic scattering of an incoming particle on the gas nuclei. This opens an avenue to use the SPC for the direct detection of low mass galactic Dark Matter (DM). A major advantage of the SPC as a light DM detector is the possibility to match the target mass to that of the candidate and, thus, maximise the recoil energy. The NEWS-G experiment [4] employs SPCs to search for DM candidates predicted by a plethora of models in the 100 MeV/c^2 to 10 GeV/c^2 mass region with an unprecedented sensitivity.

IOP Conf. Series: Journal of Physics: Conf. Series **1029** (2018) 012006 doi:10.1088/1742-6596/1029/1/012006

2. The Spherical Proportional Counter

The Spherical Proportional Counter (SPC) is a gaseous detector that is based on the spherical geometry. The detector vessel can be pumped to a high vacuum to remove impurities then be filled with a gas mixture of choice, potentially up to high pressure. The vessel is grounded and acts as the cathode. The anode consists of a small ball, usually made from a metallic or a resistive material, placed in the center of the vessel and supported by a grounded metallic rod, through which the high voltage is applied. A schematic representation of the detector is shown in Figure 1.

The electric field intensity in the detector rapidly varies with the distance from the centre as $1/r^2$. The difference in the intensity of the electric field, from the outer to the inner radii, divides the detector volume into the drift and the amplification regions. Primary ionization electrons move towards the anode with a drift time that varies from μs to ms depending on the gas mixture and pressure. When these electrons reach a distance of a few millimeters from the anode, the avalanche starts due to the intense electric field. The pulse shape of the signal depends on the spatial distribution of the charge and the distance of the interaction from the anode, because these factors define the spread in arrival time of charges and so the shape of the pulse. A complete description of the SPC, including operation principles, characteristics and capabilities, is given in Ref. [2].

The main advantages of the SPC that make it appropriate for a number of applications are: a) it can be built solely out of radio-pure materials; b) it can sustain high pressure; and c) it allows operation in sealed mode for long periods of time. Below some of these are discussed in more detail.

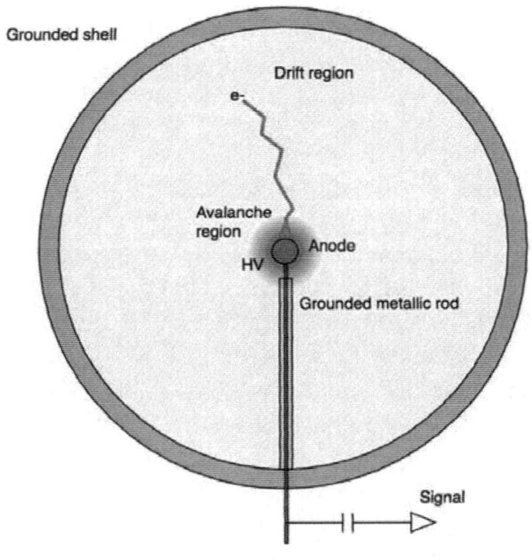

Figure 1. A schematic of the Spherical Proportional Counter and the detection principle.

An important advantage offered by a spherical detector is that large volumes can be realised with very low electronic noise. This is a direct result of the small intrinsic capacitance of the detector [5]. In contrast to the parallel plate and cylindrical geometries, where the capacitance scales proportionally to the surface and the length of the detector, respectively, the capacitance in the case of a spherical detector is given by

$$C = 4\pi\epsilon\epsilon_0 \frac{r_c r_a}{r_c - r_a} \tag{1}$$

where r_c is the radius of the outer detector shell and r_a is the radius of the central anode electrode. For a large volume SPC one has $r_c \gg r_a$, and the capacitance tends to $4\pi\epsilon\epsilon_0 r_a$. As the capacitance is independent of r_c, large volume SPCs can be constructed without a deterioration on the noise characteristics.

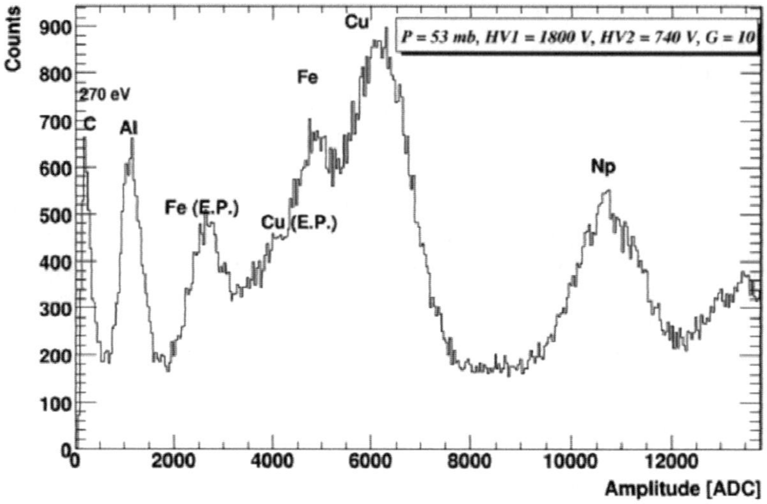

Figure 2. Fluorescence peaks observed from the ^{241}Am radioactive source through aluminum and polypropylene foils. On the left the Carbon (270 eV) line is shown, followed by the Aluminum peak (1.45 keV), the escape peak (E.P.) of Iron in Argon (3.3 keV), the escape peak of Copper in Argon (5 keV), the Iron line (6.4 keV), the Copper line (8 keV) and the Neptunium line (13.93 keV). From Ref. [6, 7]

The SPC can detect low-energy X-rays with a very good energy resolution, better than 10% (RMS) at 5.9 keV. An example of low-energy X-ray detection is presented in Figure 2, where the fluorescence lines of a series of elements with energies below approximately 15 keV are measured and resolved by a SPC with a diameter of 165 cm. The capability of an SPC to detect single electrons has been demonstrated, as shown in Figure 3. The detection energy threshold is only limited by the W-value and the primary ionization statistics of the gas mixture used.

The detection energy range varies from a few tens of eV, thanks to operation with high amplification, to tens of MeV, thanks to the large volume and high pressure operation. This permits the detection of low energy photons, electrons, α particles, neutrons and heavy ions.

Finally, the SPC can readily be fiducialized [8]. The arrival time dispersion of the primary electrons liberated in the detection volume by ionizing radiation results in pulses with different shape characteristics depending on the nature of the interaction. Events with extended energy depositions can be distinguished from those with point-like energy deposition, by applying pulse-shape analysis. In this case a SPC with a diameter of 30 cm is irradiated with an ^{241}Am-^{9}Be source [9]. The recoils induced by the scattering of fast neutrons from the source have a range substantially shorter than the range of the cosmic muons transversing the entire detector. This difference in their interaction with the detector is depicted in the width, the full width at half maximum of the pulse height distribution, of the recorded pulses. Other differences in the characteristics of the interaction can be reflected in other pulse shape parameter such as the difference in the interaction radius which is reflected in different arrival times and, thus on the rise time of the pulse. Such parameters allow the definition of the fiducial region of the detector volume and provide the possibility to reject the background which usually originates from the

IOP Conf. Series: Journal of Physics: Conf. Series **1029** (2018) 012006 doi:10.1088/1742-6596/1029/1/012006

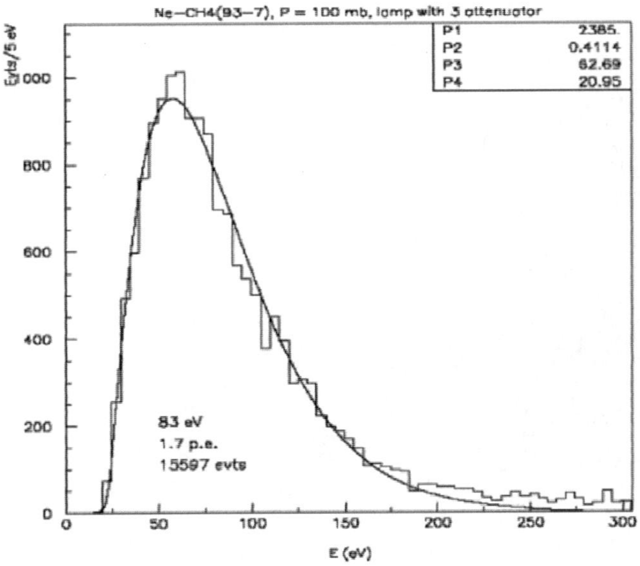

Figure 3. Energy deposition spectrum due to photo-electron extraction from the Copper shell of a SPC by irradiating the detector walls through a window with a UV lamp after attenuating the pulse to ensure single electron detection [6, 7].

outer layers of the detector, based only on the ionisation signal.

3. Applications

In this section some of the applications of the SPC are presented.

3.1. Dark Matter Search

The detection efficiency of SPCs for sub-keV nuclear recoils motivated the construction of an apparatus for the direct measurement of low mass galactic DM. These hypothetical particles are predicted by a series of models, the search for which has become much more interesting after the recent bounds set by the LHC and direct detection search for WIMPs of mass over 10 GeV/c^2. Still the direct detection of DM particles below 10 GeV/c^2 is challenging because of the low energy threshold required and the effect of the quenching factor [9] that limit dramatically the efficiency of detectors using heavy elements (Ar, Xe, Ge, etc). For this reason, large, state-of-the-art experiments have a limited sensitivity to candidate masses below a few GeV/c^2.

The NEWS-G experiment aspires to extend the search for low-mass DM candidates predicted by a plethora of models in the 100 MeV/c^2 - 10 GeV/c^2 with an unprecedented sensitivity, employing spherical gaseous detectors specifically optimised for this search. In this versatile detector different gas mixtures can be used as targets, particularly light elements, such as H, He and Ne. These allow for an optimisation of the momentum transfers for interacting low-mass particles which, together with a more favourable quenching factor, make nuclear recoils induced by sub-GeV DM candidates measurable.

SEDINE is a 60 cm in diameter detector already installed and operating at the Laboratoire Souterrain de Modane (LSM), France. The underground laboratory offers approximately 4800 m equivalent water depth to shield from cosmic radiation. SEDINE is further shielded with layers of copper, lead and polyethylene to further reduce background counting rates. The first results of this detector with a Ne gas target were recently published [10].

IOP Conf. Series: Journal of Physics: Conf. Series **1029** (2018) 012006 doi:10.1088/1742-6596/1029/1/012006

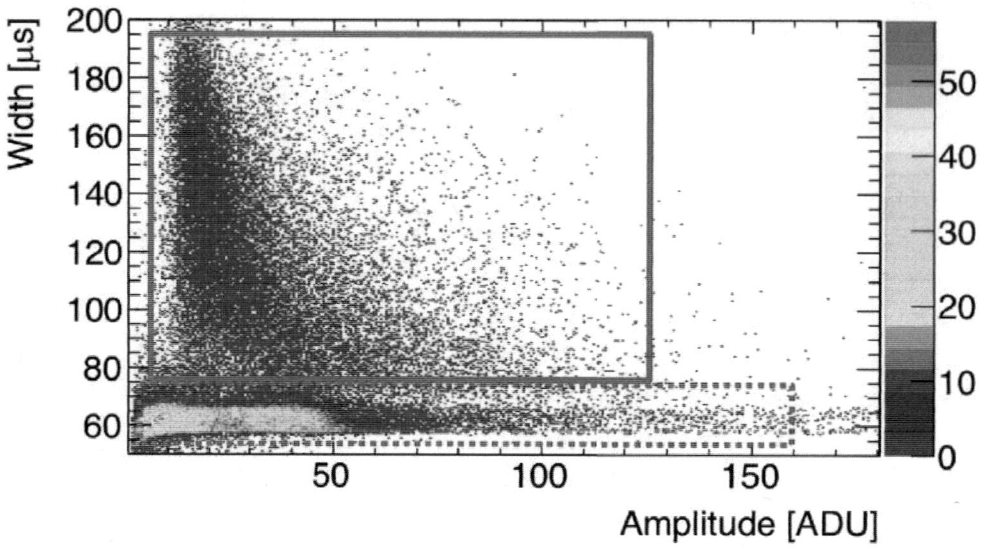

Figure 4. The width of the pulse versus energy deposition for data taken using an ^{241}Am-^{9}Be. Two distinct populations are observed corresponding to events with long ionisation tracks, such as those induced by cosmic muons, and to events with a point-like energy deposition, for instance neutron-induced recoils. These are shown in the solid and the dashed boxes, respectively. This distinction in the pulse width (full width at half maximum) allows discrimination of the interaction type.[9].

Data were collected over a period of 42.7 days filled with a gas mixture of Ne + CH_4 (0.7%) at 3.1 bars for a total exposure of 9.6 kg · days. Energy calibrations were performed using both an ^{37}Ar source that produces 270 eV and 2.82 keV x rays and neutrons from an Am-Be source, providing a large sample of nuclear recoils at sub-keV energies [10]. A calibration was also performed during the physics run, using the 8 keV photon from copper fluorescence.

To reject pulses originating from background events at the surface of the sphere, a trigger cut was applied to the rise time of a pulse. Figure 5(b) and Figure 5(a) shows the simulated distribution of surface and volume events in the sphere, respectively, which may be clearly distinguished down to the analysis energy threshold of 150 eV electron equivalent (eV_{ee}).

The analysis of the data collected was performed using a Boosted Decision Tree (BDT) [11] to optimise the signal to background discrimination for eight candidate DM masses between 0.5 GeV/c^2 and 16 GeV/c^2. Considering all events that are remaining after the cuts to be DM candidates, a 90% confidence level upper limit on the spin-independent WIMP-nucleon scattering cross section was derived from Poisson statistics and is presented in Figure 6 [12], along with other state-of-the-art searches. A new limit in the DM candidate mass-cross-section parameter space at 0.6 GeV/c^2 was obtained, with a cross-section of 4.4×10^{-37} cm^2 for a WIMP mass of 0.5 GeV/c^2 excluded at the 90% confidence level.

This result is the first in the low-mass Dark Matter candidate search to be produced using a gaseous detector. The detector is currently operating using a helium target. The results obtained with SEDINE are very promising for the main detector of the NEWS-G experiment. It will consist of a 140 cm diameter SPC made of C10100 copper, enclosed in compact shielding and is expected to be installed in SNOLAB, Canada, by the end of 2018. This is expected to

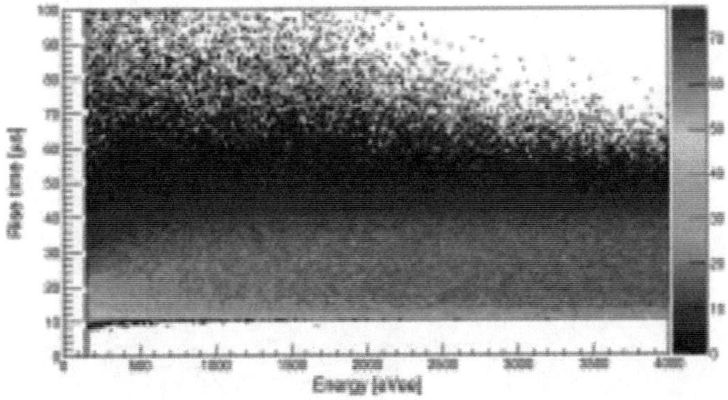

(a)

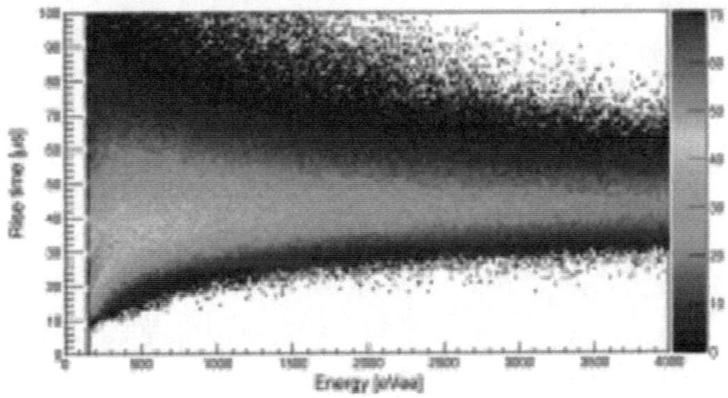

(b)

Figure 5. Rise time against energy distribution of 10^6 simulated events distributed (a) throughout the detector volume and (b) at the detector surface. Events of each type are distinguishable down to the energy threshold. [10]

have lower background rate and larger target mass, which will increase the sensitivity of the experiment by about three orders of magnitude.

3.2. Neutron Spectroscopy

The SPC technology may also be applied to direct spectroscopy of fast neutrons. Such measurements have broad interest, ranging from neutron background estimation to applications in industry. Commercially available gaseous neutron counters are generally filled with either the expensive ^{3}He or the toxic BF_3 gases [13]. Such detectors are limited by the positive Q-values of the neutron interactions which makes them sensitive to both thermal and fast neutrons. As a result, direct fast neutron spectroscopy is challenging without the use of a moderator.

The SPC provides a low cost and robust alternative to these technologies and neutron interactions have been successfully observed using ^{3}He [14]. However, the SPC may also be operated using N_2, potentially with a small amount of a quenching gas, which is cheaper than the mentioned alternatives and non-toxic. Nitrogen is sensitive to neutrons through two interactions, $^{14}N(n,p)^{14}C$ and $^{14}N(n,a)^{11}B$ which have Q-values of 625.9 keV and -158 keV, respectively. The negative Q-value of the second process allows N_2 to be used for direct fast neutron spectroscopy.

IOP Conf. Series: Journal of Physics: Conf. Series **1029** (2018) 012006 doi:10.1088/1742-6596/1029/1/012006

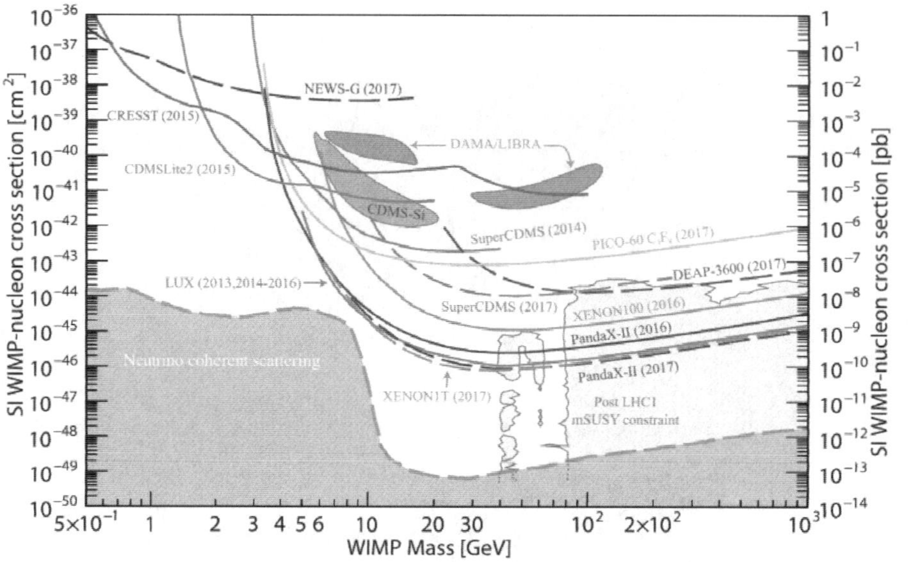

Figure 6. Experimental constraints in the WIMP-nucleon cross section vs. WIMP mass plane. The result of the NEWS-G experiment is shown as a dashed purple line, compared to the constraints and signals reported by other experiments. [10]

Measurements of the neutron spectrum of several sources were conducted using a 1.3 m copper sphere filled with N_2 with a silicon central electrode [8]. The measured rise time versus energy distribution of neutrons from a ^{252}Cf source filled with 400 mbar of N_2 is shown in Figure 7(a). A clear group of events is seen at 625.8 keV arising due to thermal neutrons interacting via the $^{14}N(n,p)^{14}C$ process, with fast neutrons appearing at higher amplitudes. The fast neutrons are detected through both interactions. Energy depositions below the peak are due to recoils induced by the elastic scattering of fast neutrons.

In Figure 7(b) the rise time versus energy distribution of atmospheric neutrons is shown following a 16-hour data-taking using 500 mbar of N_2 The 625.6 keV thermal neutron peak is clearly visible.

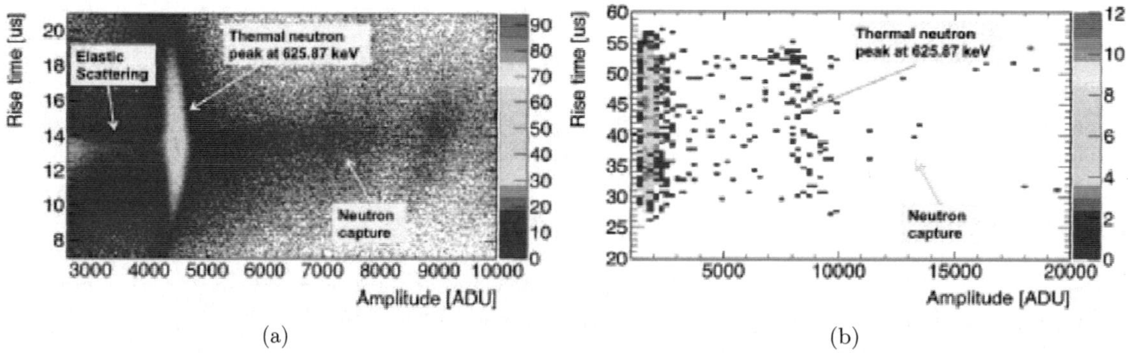

Figure 7. Measured pulse rise time versus amplitude distribution of neutrons interactions using an N_2 filled SPC, with (a) neutrons from a ^{252}Cf source and (b) using atmospheric neutron.s [8]

An exciting capability of this detector is the direct detection of fast neutrons, which has been demonstrated using both a ^{252}Cf and an Am-Be source. Fast neutrons from these sources have energies of several MeV. The spectrum of fast neutrons recorded from the ^{252}Cf source is shown in Figure 8(a) while the recoil energy spectrum from a Geant4 simulation [15] is shown in Figure 8(b). The features in the measured spectrum that are not observed in the emission energy spectrum are due to the energy dependence of the cross-section for the ^{14}N(n,a)^{11}B interaction [16].

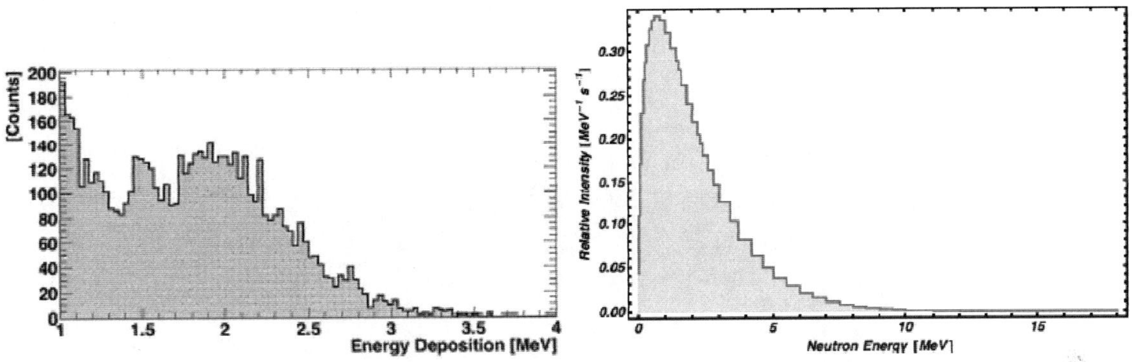

(a) (b)

Figure 8. (a) The measured energy spectrum of fast neutrons from a ^{252}Cf source. (b) The neutron energy spectrum of a ^{252}Cf source [17]. The characteristics in the measured spectrum are due to the cross-section dependence on energy. [8]

3.3. Further Applications

The use of an SPC to study neutrinoless double beta decay, $\beta\beta0\nu$, has been considered through the use of ^{136}Xe gas. Simulation studies have been performed for a copper sphere with a radius of 37 cm filled with pure ^{136}Xe at 40 bars, which is equivalent to 50 kg of gas [18]. A liquid scintillator layer surrounding this sphere is used to veto events generated in the copper or surrounding materials. For this detector mass, a background rate of 2 events per year is possible, allowing the detector to be competitive with the current generation of detectors.

The SPC technology also lends itself to neutrino detection through neutrino-nucleus coherent scattering. High pressure SPC operation, together with the cross-section enhancement due to the coherent interaction with all nucleons in the target atom, would allow detection of neutrinos from supernova [19]. For a 6 m diameter sphere filled with Xe at 10 atm with a threshold of 100 eV, one would expect between 300 and 500 events for a supernova.

It has been proposed that a global network of such relatively cheap and stable detectors could be employed to detect neutrinos from supernova explosions [19]. With many interactions in coincidence, a shielding of 100 meters of water equivalent would be sufficient to reduce the background from cosmic muons. Due to the simplicity of the detector, they could be operated by university or secondary school groups, and managed by an international scientific consortium.

4. Developments

In this section, recent developments in the geometry and materials used in the design of the central anode and support structures are discussed.

4.1. Resistive Materials

Operation at higher pressures requires a higher bias voltage on the anode to maintain both the detector gain and drift times. However, with the standard sensor design the maximum operating pressure is limited by the voltage which can be achieved without electric discharge occurring between the ball and the support rod. A solution to this problem is the use of resistive materials for a field corrector. The field corrector is mounted between the anode ball and the support rod to reduce the risk of electric discharge. This structure also provides a second electrode, whose voltage HV_2 may be used to adjust the uniformity electric field. The use of materials with a resistivity in the range $10^9\ \Omega\cdot$cm to $10^{12}\ \Omega\cdot$cm allows for the application of a bias at the surface of the material, which further reduces the risk of electrical discharge, and enables the use of larger bias voltages HV_1.

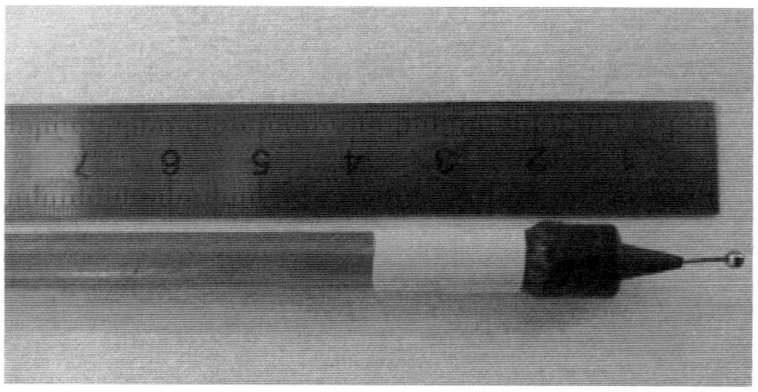

Figure 9. A sensor, showing the support rod, bakelite field corrector and anode ball.

It has been found that bakelite is ideally suited to this purpose, having a measured resistivity of the order $10^{12}\ \Omega\cdot$cm. Figure 9 shows the design of the sensor, including the bakelite field corrector. In the standard sensor arrangement, the electric field in the hemisphere containing the support rod is distorted from the ideal $1/r^2$ dependence. Such a distortion causes the same signals generated by events in the different hemispheres to have different characteristics. Through the application of a voltage HV_2 on the surface of the bakelite, the shape of the electric field may be recovered to some extent, as shown in Figure 10 .

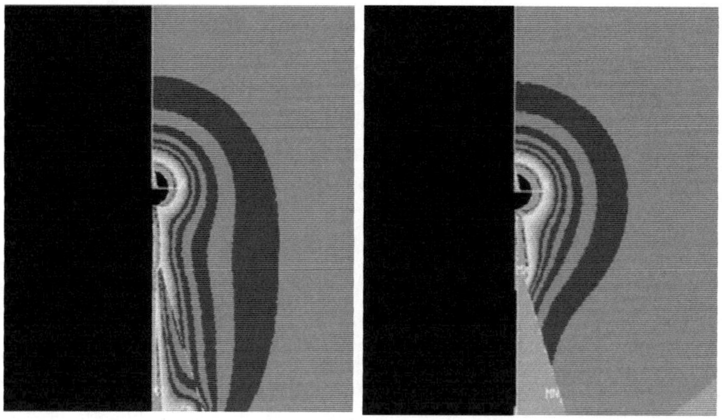

(a) (b)

Figure 10. Calculated electric field contours surrounding the anode ball of a sensor with a bakelite field corrector in the case where (a) no voltage is applied on the surface of the field corrector and (b) a voltage is applied to the surface of the field corrector.

4.2. The multiball structure: ACHINOS

In the case where the radius of the central anode r_a is very much less than the radius of the detector r_c, the radial electric field $E(r)$ in the volume may be approximated as

$$E(r) \approx \frac{V_0}{r^2} r_a \,, \tag{2}$$

where V_0 is the bias of the central anode. This proves challenging for larger detectors or operation at pressures greater than a few bar. The drift time of primary ionization particles depends on the electric field and gas pressure, however, for large radial distances the magnitude of the electric field can become very low. To compensate for this the radius of the anode ball could be increased. If a larger central anode is used, then a significantly higher bias voltage is required to maintain the same gain at the surface of the anode, which greatly increases the probability of electronic discharge to the supporting structures. Similarly, for higher pressures a smaller anode ball is required to maintain the same gain, however, this reduces the electric field at large radii.

A solution to this problem is the use of a multiball structure, called ACHINOS, to replace the single anode ball [20]. Such a structure comprises a central resistive sphere with smaller anode balls projected from this in a regular pattern and is presented in Figure 11(a). The anode balls are placed at a constant distance from the central sphere so that they lie on the surface of a virtual sphere with larger radius than central sphere. A high voltage HV_1 is applied to the individual anode balls through kapton insulated wires joining them to the central sphere. A high voltage HV_2 may also be applied on the surface of the central sphere, such that it optimises the electric field configuration. Signals may be either read out from each high voltage wire individually or by connecting all wires together. The former creates a spherical Time Projection Chamber (TPC), allowing 3D capability.

In Figure 11(b) the calculated electric field of a 2 mm anode ball sensor placed inside a shell of diameter 300 mm is compared to that of ACHINOS structures with 5, 11 and 33 anode balls, as calculated from a COMSOL simulation. The anode balls of the ACHINOS structure are distributed over a virtual sphere of 41 mm in diameter. The magnitude of the electric field at the surface of the shell is approximately 9 times higher for the 11-ball ACHINOS structure than the single ball structure.

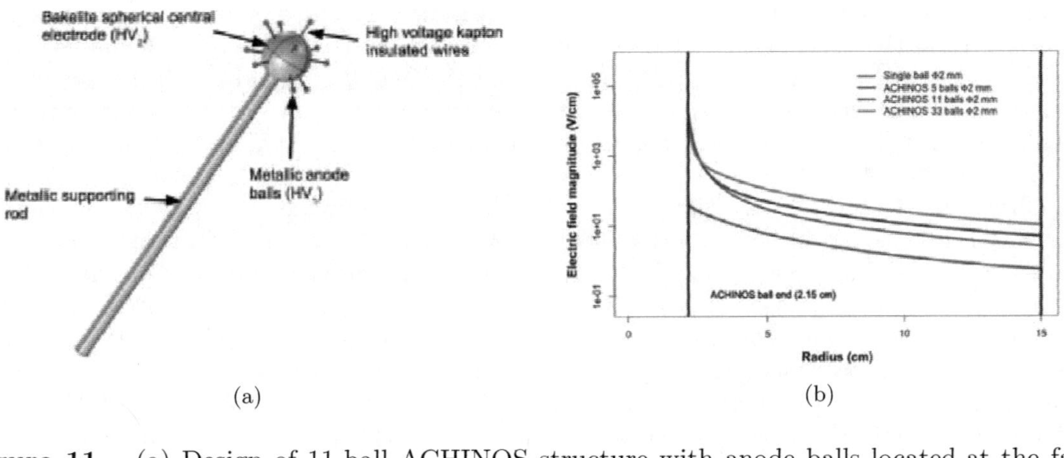

(a) (b)

Figure 11. (a) Design of 11-ball ACHINOS structure with anode balls located at the faces of a regular polyhedron (in this case a pentagonal dodecahedron). (b) The calculated Electric field as a function of radius for a standard 2 mm ball anode at HV_1 2015 V, and 5, 11 and 33 ball ACHINOS sensors, with the anode balls at HV_1 2015 V. [20]

Several prototypes of the ACHINOS structure have been constructed and tested. Figure 12 shows a comparison of the rise time versus amplitude distribution of a single 2 mm ball anode and an 11-ball ACHINOS structure with 2 mm anode balls, with the anode balls in each case being set at HV_1=2015 V. Thanks to the strong electric field at larger radii, the width of the rise time distribution of the ACHINOS sensor is 1.1 μs, much narrower than the distribution of 9.6 μs for the single ball anode.

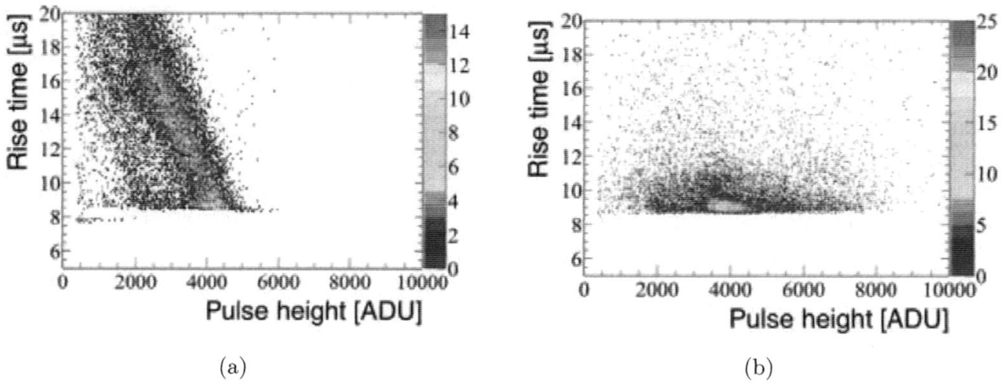

Figure 12. The rise time as a function of amplitude (a) for a standard 2 mm single ball anode at HV_1 2015 V and (b) for an 11-ball ACHINOS structure with each 2 mm ball at HV_1 2015 V. [20]

5. Summary

The design and operation principles of the SPC are presented, along with demonstrations of the response to low energy X-rays and single electrons. Recent developments in the design of the support structure of the central anode improve the uniformity of the electric field around the anode and reduce the occurrence of electrical discharges. This is achieved through the use of a bakelite structure below the anode, which permits operation of the SPC at higher gas pressures. A second design of sensor using multiple anode balls situated symmetrically about a central structure goes further by allowing a larger central structure, and so increasing the magnitude of the electric field at large radii, while also maintaining the size of each individual anode spheres, ensuring that the same gain as a single anode structure may be achieved for the same applied voltage. The capability to individual read out of the anode balls would allow TPC-like operation with 3D track reconstruction.

Applications of the SPC are discussed, particularly the ability of the detector to directly detect fast neutrons and its use in direct DM searches. The first results of a direct DM search using an SPC based in LSM established a new upper limit on the spin-independent WIMP-nucleon scattering cross-section at the 90% confidence level of 4.4×10^{-37} cm^2 for a WIMP mass of 0.5 GeV/c^2. A forthcoming DM search using a helium based target gas will be a better kinematic match for low mass DM candidates, and should increase the sensitivity of the detector in the sub-GeV range. The detector with a diameter of 140 cm, to be installed at SNOLAB near the end of 2018 at SNOLAB, is expected to improve the sensitivity by three orders of magnitude.

IOP Conf. Series: Journal of Physics: Conf. Series **1029** (2018) 012006 doi:10.1088/1742-6596/1029/1/012006

References

[1] Giomataris I *et al.* 2006 *Nucl. Phys. Proc. Suppl.* **150** 208–213 [,208(2005)] (*Preprint* hep-ex/0502033)
[2] Giomataris I *et al.* 2008 *JINST* **3** P09007 (*Preprint* 0807.2802)
[3] Giomataris Y and Vergados J 2004 *Nucl. Instrum. Meth.* **A530** 330 – 358 ISSN 0168-9002 URL http://www.sciencedirect.com/science/article/pii/S016890020401071X
[4] NEWS-G homepage. Accessed 24/01/2018 URL https://news-g.org/
[5] Andriamonje S *et al.* 2009 *J. Phys. Conf. Ser.* **179** 012003
[6] Bougamont E *et al.* 2011 *J. Phys. Conf. Ser.* **309** 012023
[7] Bougamont E *et al.* 2012 *International Journal of Modern Physics* **3** 57 – 63
[8] Bougamont E *et al.* 2017 *Nucl. Instrum. Meth.* **A847** 10–14 (*Preprint* 1512.04346)
[9] Savvidis I *et al.* 2018 *Nucl. Instrum. Meth.* **A877** 220–226 (*Preprint* 1606.02146)
[10] Arnaud Q *et al.* (NEWS-G) 2018 *Astropart. Phys.* **97** 54–62 (*Preprint* 1706.04934)
[11] Hocker A *et al.* 2007 *PoS* **ACAT** 040 (*Preprint* physics/0703039)
[12] Patrignani C *et al.* (Particle Data Group) 2016 *Chin. Phys.* **C40** 100001
[13] Knoll G 2010 *Radiation Detection and Measurement* (John Wiley & Sons) ISBN 9780470131480
[14] Savvidis I *et al.* 2010 *Journal of Physics: Conference Series* **203** 012030
[15] Agostinelli S *et al.* 2003 *Nuclear Instruments and Methods in Physics Research Section A: Accelerators, Spectrometers, Detectors and Associated Equipment* **506** 250 – 303 ISSN 0168-9002 URL http://www.sciencedirect.com/science/article/pii/S0168900203013688
[16] Chadwick M *et al.* 2011 *Nuclear Data Sheets* **112** 2887–2996 ISSN 0090-3752
[17] *International Organization for Standardization, Geneva, Switzerland*
[18] Meregaglia A *et al.* 2018 *JINST* **13** P01009 (*Preprint* 1710.04536)
[19] Vergados J D and Giomataris Y 2007 *Physics of Atomic Nuclei* **70** 140–149 ISSN 1562-692X URL https://doi.org/10.1134/S1063778807010164
[20] Giganon A *et al.* 2017 *JINST* **12** P12031 (*Preprint* 1707.09254)

Institute of Physics
Dirac House, Temple Back
Bristol BS1 6BE UK

ISSN: 1742-6588
ISBN 978-1-5108-6489-4

CURRAN ASSOCIATES INC.
proceedings.com